T0075246

Wearable Brain-Computer Interfaces

This book presents a complete overview of the main EEG-based Brain-Computer Interface (BCI) paradigms and the related practical solutions for their design, prototyping, and testing. Readers will explore active, reactive, and passive BCI paradigms, with an emphasis on the operation for developing solutions, addressing the need for customization.

Readers will familiarize themselves with the main steps for the realization of low-cost wearable BCIs which include: identification of the most suitable neuro signals for a specific application; definition of the hardware, firmware, and software, with a focus on wearable, non-invasive, and low-cost solutions; development of algorithms for data processing and classification; and, lastly, experimental campaigns for the validation of the prototyped solutions. BCI systems based on electroencephalography (EEG) are investigated and a complete overview of all BCI paradigms is offered. The aim of this book is to drive the reader, from the beginning to the end, along a research-and-development process of a working BCI prototype.

This book is a guide for designers, biomedical engineers, students, biotechnologists, and those in the biomedical instrumentation field that would like to conceive, design, prototype, and test an innovative low-cost wearable EEG-based BCI.

Wearable Brain-Computer Interfaces

Prototyping EEG-Based Instruments for Monitoring and Control

Pasquale Arpaia
Antonio Esposito
Ludovica Gargiulo
Nicola Moccaldi

CRC Press is an imprint of the
Taylor & Francis Group, an **informa** business

Designed cover image: Abstract Concept Related Use Brain-Computer Interface Stock Illustration 1972278596 | Shutterstock

First edition published 2023
by CRC Press
6000 Broken Sound Parkway NW, Suite 300, Boca Raton, FL 33487-2742

and by CRC Press
4 Park Square, Milton Park, Abingdon, Oxon, OX14 4RN

CRC Press is an imprint of Taylor & Francis Group, LLC

ISBN: 978-1-032-20085-9 (hbk)
ISBN: 978-1-032-20499-4 (pbk)
ISBN: 978-1-003-26387-6 (ebk)

DOI: 10.1201/9781003263876

Typeset in SFRM1000
by KnowledgeWorks Global Ltd.

Publisher's note: This book has been prepared from camera-ready copy provided by the authors.

May this book be a free smile to the unknown reader,
like a Light flapping of a butterfly wing
devastating the plans of Darkness
in this part of the universe.

Contents

Foreword xi

Preface xiii

Acknowledgments xvii

List of Acronyms xix

List of Figures xxiii

List of Tables xxxi

Abstract xxxv

Introduction xxxvii

I Background 1

1 Electroencephalography-Based Brain-Computer Interfaces 3

1.1 HISTORY 3
1.2 TAXONOMY 5
1.3 ARCHITECTURE 8
1.3.1 Signal Acquisition 8
1.3.2 Features Extraction 11
1.3.3 Features Translation 13
1.4 NON-INVASIVE MEASUREMENT OF NEURAL PHENOMENA 15
1.4.1 Electroencephalography 16
1.4.2 Magnetoencephalography 17
1.4.3 Functional Magnetic Resonance Imaging 17
1.4.4 Functional Near-Infrared Spectroscopy 17
1.4.5 Other Techniques 18
1.5 MEASURING THE ELECTRICAL BRAIN ACTIVITY . . 19
1.5.1 Measurand Brain Signals 20
1.6 PARADIGMS: REACTIVE, PASSIVE, AND ACTIVE . . . 23

2 Design of Daily-Life Brain-Computer Interfaces 25
2.1 ACQUISITION SYSTEM 25
2.2 ELECTRODES 29
2.2.1 Electrode-Electrolyte Interface and Polarization 29
2.2.2 Electrode-Skin System 32
2.2.3 Wet and Dry Electrodes 33
2.3 CHANNEL MINIMIZATION STRATEGIES 35
2.4 CHARACTERIZATION OF LOW-COST ELECTROENCEPHALOGRAPHS 36
2.4.1 Experiments 37
2.4.2 Data Analysis 39
2.4.3 Discussion 42
2.5 CYBERSECURITY AND PRIVACY ISSUES 46

II Reactive Brain-Computer Interfaces 49

3 Fundamentals 51
3.1 DETECTION OF STEADY-STATE VISUALLY EVOKED POTENTIALS 51
3.1.1 Physiological Basis 51
3.1.2 Measurement Setup 52
3.2 STATEMENT OF THE METROLOGICAL PROBLEM . . 53
3.2.1 Requirements 53
3.2.2 Implementations 54
3.2.3 Perspectives 55
3.2.4 Metrological Considerations 55
3.2.5 Signal Quality 57
3.2.6 Smart Glasses Characterization 61

4 SSVEP-Based Instrumentation 63
4.1 DESIGN 63
4.2 PROTOTYPE 64
4.2.1 Augmented Reality Glasses 65
4.2.2 Single-Channel Electroencephalography 66
4.2.3 Data Processing 67
4.3 PERFORMANCE 71
4.3.1 Frequency Domain 71
4.3.2 Time Domain 81
4.3.3 Response Time 86
4.3.4 Comparison with Literature 86

5 Case Studies 89
5.1 INDUSTRIAL MAINTENANCE 89
5.2 CIVIL ENGINEERING 90
5.3 ROBOTIC REHABILITATION 93

III Passive Brain-Computer Interfaces **99**

6 Fundamentals **101**
6.1 MEASURANDS . . . 101
6.1.1 Attention in Rehabilitation . . . 102
6.1.2 Emotional Valence and Human-Machine Interaction . 103
6.1.3 Work-Related Stress . . . 104
6.1.4 Engagement in Learning and Rehabilitation . . . 105
6.2 STATE OF THE ART OF EEG MARKER IN PASSIVE BCI 106
6.2.1 Attention Detection . . . 107
6.2.2 Emotional Valence Assessment . . . 107
6.2.3 Stress Monitoring . . . 111
6.2.4 Engagement Recognition . . . 115
6.3 STATEMENT OF THE METROLOGICAL PROBLEM . . 116

7 EEG-Based Monitoring Instrumentation **119**
7.1 DESIGN . . . 119
7.1.1 Basic Ideas . . . 119
7.1.2 Architecture . . . 120
7.2 PROTOTYPE . . . 124
7.2.1 Signal Pre-Processing and Features Extraction . . . 124
7.2.2 Classification . . . 126

8 Case Studies **129**
8.1 ATTENTION MONITORING IN NEUROMOTOR REHABILITATION . . . 129
8.1.1 Data Acquisition . . . 129
8.1.2 Processing . . . 131
8.1.3 Results and Discussion . . . 132
8.2 EMOTION DETECTION FOR NEURO-MARKETING . . 133
8.2.1 Data Acquisition . . . 133
8.2.2 Processing . . . 136
8.2.3 Results and Discussion . . . 137
8.3 STRESS ASSESSMENT IN HUMAN-ROBOT INTERACTION . . . 139
8.3.1 Data Acquisition . . . 141
8.3.2 Processing . . . 142
8.3.3 Results and Discussion . . . 143
8.4 ENGAGEMENT DETECTION IN LEARNING . . . 145
8.4.1 Data Acquisition . . . 147
8.4.2 Processing . . . 148
8.4.3 Results and Discussion . . . 149
8.5 ENGAGEMENT DETECTION IN REHABILITATION . . 152
8.5.1 Data Acquisition . . . 152
8.5.2 Processing . . . 153

8.5.3 Results and Discussion 155

IV Active Brain-Computer Interfaces 159

9 Fundamentals 161

9.1 STATEMENT OF THE METROLOGICAL PROBLEM . . 161

9.1.1 Motor Imagery 162

9.1.2 Neurofeedback in Motor Imagery 162

9.2 DETECTION OF EVENT-RELATED (DE)SYNCHRONIZATION 164

9.2.1 Neurophysiology of Motor Imagery 165

9.2.2 Time Course of Event-Related Patterns 166

10 Motor Imagery-Based Instrumentation 171

10.1 DESIGN 171

10.2 PROTOTYPE 172

10.2.1 Filter Bank 173

10.2.2 Spatial Filtering 174

10.2.3 Features Selection 175

10.2.4 Classification of Mental Tasks 177

10.3 PERFORMANCE 178

10.3.1 Benchmark Datasets 178

10.3.2 Testing the Feature Extraction Algorithm 179

10.3.3 ERDS Detection 181

11 Case Studies 187

11.1 WEARABLE SYSTEM FOR CONTROL APPLICATIONS 187

11.1.1 Experiments 187

11.1.2 Discussion 191

11.2 USER-MACHINE CO-ADAPTATION IN MOTOR REHABILITATION 197

11.2.1 Experiments 200

11.2.2 Discussion 204

Bibliography 209

Index 247

Foreword

Science fiction has been speculating on extending mental abilities, e.g., by proposing agents with mind reading or telekinesis capabilities. However, the possibility to communicate and control tools via brain activity is not fiction anymore. Nowadays, Brain-Computer Interfaces (BCIs) are enabling human-machine interaction by relying on the measurement and decoding of the brain activity of their users. Research in this field discloses more and more possibilities for people with neurological disease and for neurorehabilitation, but also in industry, entertainment, and gaming. It also provides more insights on brain functioning. The interest in this technology is growing, and as new companies invest in such neural interfaces, these are getting closer to daily life usage.

Pasquale Arpaia has been investigating Brain-Computer Interfaces since 2014 and, together with Nicola Moccaldi, the first approaches were exploiting the measurement of motor imagery or visually evoked potentials. Since the very beginning, the usage of few dry electrodes, low-cost instrumentation, and lightweight processing were key constraints. Then, in 2018, Antonio Esposito joined the team and the "Augmented Reality for Health Monitoring laboratory" (ARHeMlab) was founded with BCI as a pillar. At that time, a wearable BCI based on steady-state visually evoked potentials was developed and validated in an industry 4.0 scenario. This control instrument became the topic of what today is a highly cited scientific paper. Moreover, its applications are being extended to several fields, such as civil engineering and healthcare. Other than that, since 2019 the ARHeMlab group has invested new energy in motor imagery paradigms, and it also opened a new branch of investigations in the measurement of stress, emotions, as well as attention, engagement, and gait analysis. These last topics are investigated specifically by Ludovica Gargiulo, a new member of this expanding laboratory.

The book presents a fascinating journey through the research and development of wearable BCIs at ARHeMlab. The first point of specific interest is that few groups have been investigating the three paradigms for monitoring and control, namely reactive, passive, and active BCIs. So, the book provides a comprehensive overview in this sense. Then, the book presents practical solutions that can be customized depending on specific needs. Hence, it can serve as a useful reference for graduate students, researchers, and industry professionals. Last, but not least, the book was written from the point of view of a metrologist. Today, it is still rare to treat BCI in applied metrology, while the science of measurement can provide great contributions to make affordable both acquiring and processing brain signals. Overall, the book is a valuable starting point for understanding BCIs, but it may also give a new operating perspective to those already working in the field. So, enjoy the reading!

Pim Haselager
Professor of Societal Impact of Artificial Intelligence
Donders Institute for Brain, Cognition and Behaviour
Radboud University
Nijmegen, The Netherlands

Preface

This book is born from a dream, shared by all the co-authors.

I have always thought that the dream is one of the fundamental ingredients not only of research, but of the life as a whole. A dream turned into a project, and the project into winning actions, through the goodwill of everyday life. Aristotle taught us that *we are what we repeatedly do, so excellence is not an act, but a habit.*

Today everything has changed since the euphoria of the mythical 1960s in Europe. Our fathers rebuilt the abundance of Leopardi's "magnificent fortunes and progressions." Fifteen years after the war lost for everyone, when the Europe as a whole was reduced to a pile of rubbles. Between '68 and '77, we teens could remake everything, the world and its laws, by taking back our lives. We used to see even the gypsies happy. Today, on the other hand, we are in a historical period in which society gives young people harsh depressive messages. Technology research is relegated to the last rung of the STEM professional ladder. Finding young people willing to give up industrial salaries for the love of research is increasingly difficult.

When I started my job in Naples in 2014 at Federico II, I was returning from Geneva and CERN, after the drunken happiness of the post Higgs. Notoriety had also come: people remembered me in Italy, exiled abroad, because CERN's engineering and technology had also been instrumental in the Nobel Prize. Going back to Naples was crazy. But man needs new challenges to make sense of life. For almost two years I wandered those empty labs in Naples looking for meaning, a Sense. A new research activity had to be planted in Naples, humbly starting over from scratch, first and foremost for me as well. Leaving the easy and popular road of measurements for Big Physics, for cosmology and cosmogony Research, after the opening Plenary Keynote at the most important world conference in my field (IEEE I2MTC 2014, Montevideo), was not easy. After the communion with my research group at CERN, and with my friends from the Italian community in Geneva, the loneliness in a complicated and difficult town, in a territory economically hostile to research, hit hard by the crisis. Starting with 1,000 euros a year from the university endowment was not easy, but the seemingly impossible things make life stimulating. The inner challenges are the most important, not the easy victories. In those two years of physical and mental loneliness, I went to the root of man's dreams. After the dream of the overcome challenge of the Tower of Babel with CERN LHC, the largest machine ever built by mankind, I thought to the human mind's dream of being able to move lost limbs, make those who could no longer stand and walk, move objects at a distance.

All began with an MSc student in Electronic Engineering with a final work thesis on Brain-Computer Interfaces. We used to meet at 7:30 a.m. in my office, before classes, before the agenda of seeking funding and financing, of fighting bureaucracy and non-doing. The first hurdle immediately was not technical difficulty. Rather, an unexpected door immediately opened: we were measuring steady-state visual potentials evoked in the brain by a flashing light source for navigating between menus on a smartphone. This was the feasibility *in nuce* of ARHeMlab's *BrainPhone*. Easily, with a 200-euro data acquisition system for an electroencephalography with a very-low density of active dry electrodes, highly wearable. The first challenge was convincing the young student that he could do it, that he was doing it. That he was creating a simple system, with only two dry electrodes (a Walkman's headset, in short), allowing for smartphone navigation by simply looking at flashing icons. After such a thesis work, it took us another two years to improve those early positive results, between mistaken attempts in the team and in my choices.

The student couldn't believe that he could do it, that it worked in the lab, that he had done it. The black barrier was not on the lab bench, but in his mind/heart. We should not have named the lab as Laboratory of Augmented Reality for Health Monitoring (ARHeMlab), but we should just have entitled to him, after his disbelief that he could make the dream come true. Today we are an International Research Group of about 30 people, but still the problem remains the same. To dream, not to realize. Because in action we engineers are masters.

Then came Nicola Moccaldi. I noticed a mature BSc student at Course and I assigned him the thesis to reproduce and improve a BCI research paper claiming results crazily better than ours. Nicola bravely proved that it was a hoax. He already had a BSc in Communications Science but decided to move on with the humanities. He was immediately on our side, because I liked his ability to analyze the state of the art and discern the research trend. Then, after a year spent in our electronic syringe spin-off in aesthetic medicine, he joined the BCI team definitively. Awards and an H index almost as an associate professor testify today as I saw right then.

Next was the turn of Antonio Esposito. Very young and intellectually honest to the point of self-destruction, he won the competition for a thesis at CERN with me owing to his outstanding CV of studies and evident coherent human qualities. After a year of very hard work in Geneva, his very young age stopped him, proving that perfection is fortunately tedious in this part of the universe. After a few months of Peter Pan syndrome, he focused that his way was the academic research and also joined the BCI Team permanently. His scientific excellence and his love for teaching today prove that I had got it right in this case, too.

Last but not least, here we are at Ludovica Gargiulo. Also very young and animated by a sacred determination, perfectly concealed by her physical levity. In my optimizing also the resource at ppm, I had her enter five doctoral competitions. Without a wrinkle she won three of them, and at the end she too joined Team BCI (passive). A perfect embodiment of Goethe's saying, unhurried but without pause, she grinds work lightly, as if task boulders were colored balloons. Her results after the first year of her doctoral program speak for her: res, non verba.

In a nutshell, I could say that this book is the result of what we have learned about BCIs over the years. But I would be enormously reductive and ungrateful to life: in fact, the technologies and ideas with the results set forth here are only the tip of the iceberg. Not because we have been furiously working, invaded and mutually infected by the same passion, on other research topics as well. But because we have grown and this journey has made us different. Inextinguishably. As my friend Sandra Castellani, anthropologist and writer, says: It is not important where we arrive, but how we walk. And we walked together. After the most difficult moments, however, turning around, behind us we did not find the footsteps of us all, but only those of One. And so it is to this One toward whom we tend the bows of our lives to hurl the knowledge of our readers forward.

Pasquale "Papele" Arpaia

Napoli Mergellina, on the 4th of July, 2022

Acknowledgments

The Authors warmly thank all those who, through their suggestions and work, contributed to the realization of this book.

First of all, our thanks go to the members of the Laboratory of Augmented Reality for Health Monitoring (ARHeMLab) for their continuous support in the experiments and in revising some book material.

In particular, the Authors would like to thank Giuseppina Affinto, Andrea Apicella, Francesco Caputo, Giusy Carleo, Sabatina Criscuolo, Anna Della Calce, Giampaolo D'Errico, Luigi Duraccio, Francesca Mancino, Andrea Pollastro, and Ersilia Vallefuoco, whose valuable suggestions helped to improve the quality and readability of this work.

Sincere thanks to Marc Gutierrez for his suggestions and continued support.

A special thanks goes to Leopoldo Angrisani for his unsparing support in launching and feeding the research line of BCI in our group and to Marco Parvis and Egidio De Benedetto for their brilliant and constant support over all these years.

The Authors thank Angela Natalizio for her support in active BCI studies, and Luigi Duraccio and Silvia Rossi for their contribution in wearable Brain–Computer Interface instrumentation for robot-based rehabilitation by augmented reality.

For the research in passive BCI, the Authors thank Andrea Apicella, Mirco Frosolone, Salvatore Giugliano, Giovanni Improta, Francesca Mancino, Giovanna Mastrati, Andrea Pollastro Roberto Prevete, Isabella Sannino, and Annarita Tedesco.

For their outstanding contribution to the metrological characterization of BCI consumer-grade equipment, the authors thank Luca Callegaro, Alessandro Cultrera, Massimo Ortolano.

The Authors thank G. De Blasi and E. Leone, for the support in software management of robotic BCI/XR-based rehabilitation solutions, as well as Carmela Laezza for the support in data elaboration for the stress monitoring.

The Authors would like to thank A. Cioffi, Giovanna Mastrati, and Mirco Frosolone for their support during the experimental campaigns on (i) inspection in Industry 4.0, (ii) attention monitoring, and (iii) emotional valence measurement, respectively.

Part of the research was supported by (i) the Italian Ministry of University and Instruction (MIUR), through the "Excellence Department project"

(LD n. 232/2016), (ii) regione CAMPANIA through the project "Advanced Virtual Adaptive Technologies e-hEAlth" (AVATEA CUP. B13D18000130007) POR FESR CAMPANIA 2014/2020, and (iii) INPS - National Social Security Institution, through the PhD Grant "AR4ClinicSur- Augmented Reality for Clinical Surgery".

The Authors thank ab medica s.p.a. for providing the instrumentation in the experiments, whose support is gratefully acknowledged.

The Authors thank Francesco Isgrò and Roberto Prevete for stimulating discussions on Machine Learning strategies, and Andrea Grassi for useful suggestions on human-robot interaction in the workplace.

The Authors would like to thank Leda Bilo, Paola Lanteri, Luigi Maffei, and Raffaele Maffei for their supervision on neurophysiological themes.

List of Acronyms

EEG Electroencephalography

BCI Brain-Computer Interface

IEEE Institute of Electrical and Electronic Engineers

ITR Information Transfer Rate

ECoG Electrocorticogram

SMRs Sensorimotor rhythms

VEPs Visually-Evoked Potentials

SSVEP Steady-State Visually-Evoked Potential

LDA Linear Discriminant Analysis

kNN k-Nearest Neighbors

SVM Support Vector Machine

RF Random Forest

MEG Magnetoencephalography

MRI Magnetic Resonance Imaging

CT Computed Tomography

PET Positron Emission Tomography

fMRI Functional Magnetic Resonance Imaging

EMG Electromyography

ANN Artificial Neural Network

CNN Convolutional Neural Network

CV Cross Validation

SAM Self-Assessment Manikin

RACNN Regional-Asymmetric Convolutional Neural Network

fNIRS Functional Near-Infrared Spectroscopy

SPECT Single-Photon Emission Computed Tomography

DGCCN Dynamical Graph Convolutional Neural Network

EOG Electroculography

ECG Electrocardiogram

EPs Evoked Potentials

ERD Event-Related Desynchronization

ERS Event-Related Synchronization

UART Universal Asynchronous Receiver-Transmitter

USB Universal Serial Bus

MI Motor Imagery

IQR Interquartile Range

NBPW Naive Bayesian Parzen Window

CSP Common Spatial Pattern

DRL Drive Right Leg

ADC Analog-to-Digital Conversion

IVD Inductive Voltage Divider

RVD Resistive Voltage Divider

ReCEPL Research Centre of European Private Law

FBCSP Filter Bank Common Spatial Pattern

AI Artificial Intelligence

AR Augmented Reality

VR Virtual Reality

MR Mixed Reality

XR Extended Reality

FFT Fast Fourier Transform

PSD Power Spectral Density

SVC Support Vector Classifier

NB Naive Bayes

FIR Finite Impulse Response

CCA Canonical Correlation Analysis

ICA Independent Component Analysis

PCA Principal Component Analysis

ASR Artifact Subspace Reconstruction

NF Neurofeedback

ANOVA One-Way Analysis Of Variance

IAPS International Affective Picture System

GAPED Geneva Affective Picture Database

PPG Photoplethysmogram

RACNN Regional-Asymmetric Convolutional Neural Network

MLF-CapsNet Multi-Level Features guided Capsule Network

SEED SJTU Emotion EEG Dataset

CPT Continuous Performance Test

List of Figures

1.1 A classic non-invasive technique for measuring brain activity, the EEG. 6
1.2 Representation of an invasive intracortical sensing [21]. . . . 6
1.3 ECoG as an example of partially invasive measurement [20]. 6
1.4 General architecture of a brain-computer interface. 8
1.5 EEG electrodes locations according to the international standard framework 10-20 [30]; the positions *Fpz* and *Oz* are highlighted in black at the scalp frontal region and at the occipital region, respectively [31]. 10
1.6 Feature extraction for maximizing separability of signals from two different classes: representation in frequency (a) and feature (b) domains (a.u.: arbitrary unit). 12
1.7 Linear and non-linear classification of two different datasets represented with two features. The classifier boundaries are in black, the two possible classes are in red and blue: (a) linear classifier used for linearly separable classes, (b) non-linear classifiers attempting to separate linearly separable classes, (c) linear classifier attempting to separate non-linearly separable classes, and (d) non-linear classifier used for non-linearly separable classes [33]. 15

2.1 The Olimex EEG-SMT acquisition board with two active dry electrodes connected to channel 1 (CH^+, CH^-) for a single-channel differential acquisition and a passive dry electrode connected to “drive right leg” (DRL) reference terminal [61]. 26
2.2 The EEG cap Helmate by ab medica with ten dry electrodes in different shapes and Bluetooth connection to ensure high wearability and portability: (a) front view, (b) side view, and (c) electrodes [62]. 27
2.3 The wearable, high resolution, 14-channel acquisition system Emotiv Epoc+. [64] 28
2.4 The FlexEEG system by Neuroconcise Ltd with seven electrodes needing conductive gel. Bluetooth connectivity ensures portability other than wearability: (a) flexible board and electrode array, (b) sensorimotor electrode array, and (c) star-shaped electrodes [65]. 30

2.5 Electrode-electrolyte interface; the electrode consists of metal C atoms, while the electrolyte solution contains cations of the same metal C+ and A anions. I is the current flow. 31
2.6 Equivalent electrode circuit. 32
2.7 Equivalent circuit of the wet electrode-skin system [68]. E_{hc} is the half-cell potential, C_d is the capacitance of the capacitor relative to the electrical double layer, R_d is the charge transition resistance at the interface, and R_s is the resistance associated with the electrolyte. 33
2.8 Circuit model of the electrode-skin system for wet (a) and dry (b) electrodes [69]. E_{hc} is the half-cell potential, C_{el} is the capacitance of the capacitor relative to the electrical double layer, R_{el} is the charge transition resistance at the interface, and R_g is the resistance associated with the presence of the gel. The parallels $C_{ep}//R_{ep}$ and $C_{sw}//R_{sw}$ represent the impedance offered by the epidermis and by the sweat ducts, respectively, to the current flow. R_d represent the resistance offered by the dermis and subcutaneous layer. E_{ep} is the potential difference due to the different concentration of ions on either side of the membrane. E_{sw} represents the potential difference between the interior of the sweat ducts and dermis. . 34
2.9 Signal conditioning circuit being part of the Olimex EEG-SMT acquisition board. The connections to the CH1 pins and the DRL are shown, as well as the input ADC line. Redrawn and adapted from the datasheet of e Olimex EEG-SMT biofeedback board [91]. 38
2.10 Coaxial schematic diagram of the EEG calibration setup realized at the "Instituto Nazionale di Ricerca Metrologica". . . 39
2.11 Experimental measurement setup for the EEG calibration: (a) instruments connected according to the coaxial scheme of Fig. 2.10, and (b) detail of the connection to the electrodes. 40
2.12 Linearity of the EEG-SMT electroencephalograph measured at 20 Hz (inset: sine wave acquired with for $V_{CAL,rms}$ = 100 µV) [97]. 42
2.13 Magnitude (gain) error of the EEG-SMT electroencephalograph measured at frequencies up to 100 Hz [97]. 43
2.14 Analog conditioning circuitry of the Olimex EEG-SMT board replicated in LTspice for executing a Monte Carlo analysis. . 44
2.15 Simulated gain error obtained with a Monte Carlo analysis (gray line and shaded areas) and compared to the measured gain error (black line with bars). The different gray levels correspond to three different probabilities of occurrence for simulated values [97]. 45
2.16 Relative error between the measured frequencies and the generator ones in percentage. 47

2.17 Brain hacking, damages and countermeasures in different BCI use cases. 48

3.1 Wearable BCI system based on SSVEP: (a) the BT200 AR glasses for stimuli generation, (b) the human user, and (c) the Olimex EEG-SMT data acquisition board with dry electrodes [97]. 57
3.2 Comparison between simulated nominal amplitude spectrum of the visual stimuli (a) and measured EEG spectrum corresponding to visual stimulation (b). 58
3.3 A typical electroencephalographic signal recorded through dry electrodes during experiments with SSVEP. Artifacts related to eye-blinks are clearly detectable by the pronounced valleys in the recording. 59

4.1 Architecture of the AR-BCI system based on steady-state visually evoked potentials [31]. 64
4.2 FIR Filter Amplitude frequency response [140]. 69
4.3 Detection of voluntary and involuntary eye blinks [142]. . . . 70
4.4 An example of a signal measured with the Olimex EEG-SMT after placing all the dry electrodes for SSVEP detection. Note that some artifacts related to eye-blinks are present. 73
4.5 Scatter plot of EEG signals in the PSD features domain, associated with 10 s-long stimulation and for a different number of simultaneous stimuli. Both axes are in logarithmic scale. (a) 1 stimulus and (b) 2 stimuli. 75
4.6 Scatter plot of EEG signals in the PSD features domain, associated with 2 s-long stimulation and for a different number of simultaneous stimuli. Both axes are in logarithmic scale. “1 stimulus” (a) and “2 stimuli” (b). 76
4.7 Mean classification versus SSVEP stimulation time (almost coincident with system latency). The standard deviation of the mean is also reported as a shaded area. The “1 stimulus” case is compared with the “2 stimuli” one. 80
4.8 Median classification versus SSVEP stimulation time (almost coincident with system latency). The interquartile range is also reported with a shaded area as a measure of dispersion around the median. The cases “1 stimulus” and “2 stimuli” is compared. 81
4.9 Classification accuracy versus stimulation time in the four PSD features case. Mean and standard deviation (a) and median and interquartile range (b). 82

4.10 Box-plots of the Information Transfer Rate (ITR) as a function of SSVEP stimulation time (almost coincident with system latency). The case "1 stimulus" is here considered. The median values are reported as red lines in the boxes. The few outliers are associated with subjects that have quite higher performance in comparison with the median one. 83
4.11 Trade-off between accuracy and time response highlighted in a plot vs parameters T1 and T [140]. 84
4.12 Java Application for simulating the behavior of the robot [140]. 85

5.1 AR-BCI system based on SSVEP for accessing data from a wireless sensor network [31]. 90
5.2 Android application diagram for an inspection task in industrial framework: smart transducers and related measures are selected by staring at corresponding flickering icons [31]. . . 91
5.3 Example of inspection with the wearable SSVEP-BCI: measure selection menu, with flickering white squares for options (as usual in AR glasses, background image is blurred to focus on the selection). A user wearing the SSVEP-BCI system during an emulated inspection task (a) and Simulated measure selection window with two possible choices (b) [31]. 92
5.4 AR-BCI system based on SSVEP communicating with a sensor stack through Bluetooth low-energy. 93
5.5 User interface in Android for bridge inspection during a load test. In the first activity, (a) the Smart Glasses automatically connect to multiple wireless transducers. In the second activity (b) the user visualizes environmental data. The third and fourth activities (c and d) can be accessed by selecting the data to visualize. The black background is equivalent to transparency when the application is running on Smart Glasses. . 94
5.6 Application of the wearable SSVEP-BCI in the rehabilitation of children with ADHD through a behavioral therapy, as documented in [140]. Possible control commands in interacting with the robot (a) and Finite state machine implemented for robot control (b) [140]. 95
5.7 Distinction between voluntary and involuntary eye-blinks in time domain with an empirically determined threshold. . . . 96

6.1 Mental state detection in the Industry 4.0. 102

7.1 The architecture of the system for mental state assessment. 120
7.2 The system architecture specific for distraction-detection (CSP: Common Spatial Pattern algorithm)[307]. 121
7.3 The system architecture specific for emotion-detection (CSP: Common Spatial Pattern algorithm) [307]. 122

7.4 Architecture of the real-time stress monitoring system in Cobot interaction [140]. 122
7.5 The architecture of the system for engagement assessment in rehabilitation [307]. 123
7.6 The architecture of the system for engagement assessment in learning; the white box is active only in the cross-subject case (ADC - Analog Digital Converter, CSP - Common Spatial Pattern, TCA - Transfer Component Analysis, and SVM - Support Vector Machine) [327]. 123

8.1 Visual distractor task elements based on visual Gabor mask with different orientations, equal to 90°, 60°, and 30°, respectively [307]. 130
8.2 F-Measure test results for the best performance of each classifier: Precision (black) , Recall (gray), and F1-score (white). 133
8.3 Experimental protocol [276]. 134
8.4 Oasis valence and SAM average scores of the 26 images selected for the experiments. The Oasis score intervals used to extract polarized images are identified by dotted lines [276]. 135
8.5 Bland-Altman analysis on the agreement between stimuli (OASIS) and volunteers perception (SAM) [276]. 135
8.6 F1—score (White), Recall (Gray), and Precision (Black) for the best performance of each classifier - Cross-subject [276]. 139
8.7 Cumulative Explained Variance in the PCA [140]. 143
8.8 SVM data distribution in PCs space, p = 2, 92.6 % of explained variance [140]. 146
8.9 Screen shots from the CPT game [327]. At the beginning of the game, the cross starts to run away from the center of the black circumference (a). The user goal is to bring the cross back to the center by using the mouse. At the end of each trial, the score indicates the percentage time spent by the cross inside the circumference (b). 148
8.10 Within-subject performances of the compared processing techniques in (a) cognitive engagement and (b) emotional engagement detection. Each bar describes the average accuracy over all the subjects [327]. 149
8.11 Filter Bank impact on the class (red and blue points) separability. t-SNE-based features plot of five subjects randomly sampled (first row: without Filter Bank; second row: with Filter Bank) [327]. 150
8.12 A comparison using t-SNE of the FBCSP data first (a) and after (b) removing the average value of each subject, in the cross-subject approach [327]. 151
8.13 Neuromotor rehabilitation session [307]. 153

8.14 The experimental paradigm: only on the first day, a training phase is implemented [307]. 154
8.15 Cognitive engagement balanced accuracies for each subject based on KMeansSMOTE oversampling technique. Classifier performances are reported [307]. 156
8.16 Emotional engagement balanced accuracies for each subject based on KMeansSMOTE oversampling technique. Classifier performances are reported [307]. 157

9.1 Time-frequency representation of EEG patterns associated with (a) left hand, (b) right hand, or (c) feet motor imagery, and their spatial distribution, calculated in accordance with [395] with data from subject A03T of BCI competition IV dataset 2a. 165
9.2 Calculation of ERD (a) and ERS (b) according to the power method proposed in [53]. A decrease of band power indicates ERD while an increase of band power ERS. 167
9.3 Comparison of the power method and the inter-trial variance method for calculating the ERD (a) and ERS (b). 168
9.4 ERD (a) and ERS (b) calculated according to the power method after choosing subject-specific frequencies. The ERS is still not visible, but the ERD is much more evident. . . . 170

10.1 Block diagram of a simple BCI system based on motor imagery. Neurofeedback could be eventually provided according to the signals classification, but this is not yet represented in this diagram. 171
10.2 Wearable and portable EEG acquisition system employed in prototyping a MI-BCI (a) and its electrode positioning (b). . 173
10.3 Filter Bank Common Spatial Pattern for processing EEG signals associated with motor imagery. 173
10.4 Classification accuracies for the 12 subjects from dataset 2a and dataset 3a. The six possible classes pairs are considered as well as the four classes case: cross-validation on training data (a) and hold-out method (b). 180
10.5 Time course of ERDS in alpha (a) and beta (b) bands for subject A03 (session T) from BCI competition IV dataset 2a. 182
10.6 Analysis of MI-related phenomena for subject A03 (session T) from BCI competition IV dataset 2a: time course of classification accuracy (a) and mostly selected features (b). 183
10.7 Time course of ERDS in alpha (a) and beta bands (b) for subject A05 (session T) from BCI competition IV dataset 2a. 183
10.8 Analysis of MI-related phenomena for subject A05 (session T) from BCI competition IV dataset 2a:time course of classification accuracy (a) and mostly selected features (b). 184

10.9 Time course of ERDS in alpha (a) and beta (b) for subject A02 (session T) from BCI competition IV dataset 2a. 184
10.10 Analysis of MI-related phenomena for subject A02 (session T) from BCI competition IV dataset 2a: (a) time course of classification accuracy, and (b) mostly selected features. . . . 185

11.1 Results of motor imagery classification with single channel data from dataset 2a of BCI competition IV. The six possible pairs of classes are considered for the nine subjects: left hand vs right hand (blue), left hand vs feet (red), left hand vs tongue (yellow), right hand vs feet (magenta), right hand vs tongue (green), and feet vs tongue (cyan). 189
11.2 Representation of the channel selection algorithm exploiting the FBCSP approach for the classification of EEG signals. . 190
11.3 Classification performance (cross-validation) for progressively selected channels and for each pair of classes: left hand vs right hand (a), left hand vs feet (b), left hand vs tongue (c), right hand vs feet (d), right hand vs tongue (e), and feet vs tongue (f). 193
11.4 Classification performance (cross-validation) for progressively selected channels in the four-class problem. 194
11.5 Mean classification performance obtained to validate the respective sequence of the channel selection procedure in the binary classification cases for the six possible pairs of classes: left hand vs right hand (a), left hand vs feet (b), left hand vs tongue (c), right hand vs feet (d), right hand vs tongue (e), and and feet vs tongue (f). 195
11.6 Mean classification performance obtained to validate the respective sequence of the channel selection procedure in the four-classes case. 196
11.7 Classification performance in validating the channel sequences found on dataset 3a: right hand versus tongue (a) and four classes (b). 196
11.8 Most predictive information on the scalp for each pair of classes: left hand vs right hand (a), left hand vs feet (b), left hand vs tongue (c), right hand vs feet (d), right hand vs tongue (e), and feet vs tongue (f). 197
11.9 Visual feedback consisting of a mentally-controlled virtual ball [422]. 198
11.10 Wearable and portable haptic suit with 40 vibration motors [422]. 199
11.11 Timing diagram of a single trial in the BCI experiment with neurofeedback. 200
11.12 Blocks and sub-blocks of the neurofeedback protocol employed in the closed-loop motor-imagery-BCI. 201

11.13 An example of the permutation test for the subject SB1 executed on the time course of classification accuracy in different experimental conditions. The green curves correspond to the accuracy calculated with true labels and the black curves correspond to the accuracy obtained with permuted labels. . . 205
11.14 Time varying decoding accuracy associated with motor imagery. The classification accuracy of subject SB1 is compared to the one of a trained subject (A03 from BCI competition IV): no feedback (a) and with neurofeedback (b). 206

List of Tables

1.1 Summary of the main characteristics concerning neuroimaging methods. 19

4.1 Classification performance of SSVEP-related EEG signals. For each subject, the "1 stimulus" case is compared with the "2 stimuli" case, and the results of a 10 s-long stimulation are compared with a 2 s-long one. Performance is assessed with cross-validation accuracy and its associated standard deviation over 4-folds. The mean accuracy among all subjects is reported too, as well as the accuracy obtained by considering all subjects together (row "all"). The SVM classifier considers two PSD features. 78
4.2 Comparison of classification performance for a SVM classifier considering two PSD features (2D SVM) and one considering four PSD features (4D SVM) of SSVEP-related EEG data. The mean cross-validation accuracies and their associated standard deviations are reported for the "1 stimulus" and the "2 stimuli" case, and a 10 s-long stimulation is compared with a 2 s-long one. 79
4.3 Accuracy (%) of SSVEP detection algorithm for different time windows T and threshold values T1 [140]. 83
4.4 Time response (s) of SSVEP detection algorithm for different time windows T and threshold values T1 [140]. 84
4.5 Performance of SSVEP/Eye blink integrated detection algorithm for T=0.5 s, $T1$=0.44, and eye blink threshold equal to 0 [140]. 85
4.6 ITR (bit/min) of SSVEP/Eye blink detection algorithm at varying time windows T and thresholds $T1$ [140]. 86

5.1 Clinical case study: Performance of SSVEP/Eye blink integrated detection algorithm for T=0.5 s, $T1$=0.44, and eye blink threshold value equal to 0. 97

6.1 Studies on emotion recognition classified according to the employed datasets (i.e., SEED, DEAP, and DREAMER), stimuli (v="video", p="picture", m="memories"), task (i="implicit", e="explicit", n.a.="not available"), #channels, #participants, #classes, classifiers, and accuracies (n.a.="not available") [276]. 112
6.2 State of art of stress classification [140]. 114

8.1 Data-set composition [307]. 131
8.2 Hyperparameters considered in classifier optimization and their variation ranges [307]. 132
8.3 Classifier optimized hyperparameters and variation range [276]. 137
8.4 Percentage accuracy (mean and standard deviation) considering a priori knowledge i.e., Asymmetry—Within-subject (Within) & Cross-subject (Cross) [276]. 138
8.5 Percentage accuracy (mean and standard deviation) without considering a priori knowledge i.e., Asymmetry—Within-subject [276]. 138
8.6 Percentage accuracy (mean and standard deviation) without considering a priori knowledge i.e., Asymmetry—Cross-subject [276]. 138
8.7 Studies on emotion recognition classified according to metrological approach, number of channels, and accuracy (n.a. = "not available", ✓ = "the property is verified". "Only for the first line", ✓ = "Measurement") [276]. 140
8.8 Stress index distribution (descending sort) [140]. 142
8.9 Classifier optimized hyperparameters and range of variation [140]. 143
8.10 Accuracy (mean and uncertainty percentage) in Original Data (O.D.) and Principal Components Hyperplanes [140]. 144
8.11 Classifiers accuracy (mean and uncertainty percentage) in Original Data (O.D.) and Principal Components Hyperplanes [140]. 144
8.12 F-measure test results for SVM (mean and uncertainty percentage) [140]. 145
8.13 Accuracy (mean and uncertainty%) in Original Data (O.D.) and Principal Components Hyperplanes at varying amplitude of random Gaussian noise [140]. 145
8.14 Accuracy (mean and uncertainty) in Original Data (O.D.) and Principal Components Hyperplanes at varying amplitude of homogeneous noise (%) 146
8.15 Within-subject classification accuracies (%) using the *Engagement Index* [364] for cognitive engagement classifications [327]. 149

8.16 Within-subject experimental results. Accuracies (%) are reported on data pre-processed using Filter Bank and CSP for both cognitive and emotional engagement classifications. The best performance average values are highlighted in bold [327]. 150
8.17 Cross-subject experimental results using FBCSP followed by TCA. Accuracies (%) are reported with and without for-subject average removal for cognitive engagement and emotional engagement detection. The best performance values are highlighted in bold [327]. 152
8.18 Oversampling methods, optimized hyperparameters, and variation ranges [307]. 154
8.19 Classifiers, optimized hyperparameters, and variation ranges [307]. 155
8.20 Overall mean of the intra-individual performances on cognitive engagement using three different classifiers: the BA (%) and the MCC at varying the oversampling methods [307]. 155
8.21 Overall mean of the intra-individual performances on emotional engagement using three different classifiers: the BA (%) and the MCC at varying the oversampling methods [307]. . 156

10.1 Mean and standard deviation of the cross-validation accuracy obtained on dataset 2a with a Bayesian classifier. 179
10.2 Classification accuracies for the 12 subjects from dataset 2b and dataset3b. Subject-by-subject accuracies are reported for cross-validation (CV) and hold-out (HO). 181

11.1 Comparison between NBPW, SVM, and kNN classifiers for different classification problems. Mean cross-validation accuracy and associated standard deviation were calculated among nine subjects (dataset 2a) by taking into account all channels. 192
11.2 Mean and standard deviation of the classification accuracy obtained during channel selection, for both 8 and 22 channels. 193
11.3 Mean and standard deviation of the classification accuracy obtained during channel sequences validation. 195
11.4 Classification results for the preliminary experiments with four subjects [422]. 201
11.5 Questionnaire provided to the participants at each experimental session. 203

Abstract

A brain-computer interface (BCI) is a system capable of measuring and directly translating brain signals into instructions for peripheral devices of various types such as communication systems, wheelchairs, prostheses. Thus, it creates a communication channel independent of the normal neuromuscular outputs of the central nervous system.

In this book, *reactive*, *passive*, and *active* BCI paradigms with relevant case studies are analyzed. A common element of the solutions presented is the use of non-invasive, wearable, and low-cost systems to enable them to be used in daily life. Among non-invasive techniques, the most popular is electroencephalography (EEG), due to its good portability and temporal resolution.

The book begins with a minimal background on the electroencelography underlying Brain-Computer Interfaces (Chapter 1) and related technologies for the design of related wearable systems for everyday use (Chapter 2). The main focus is on data acquisition with few channels and electrodes, dry and active. Then, the description is structured along three directions.

The first direction concerns *reactive* BCI, with an external stimulus, in our case a visual one. For example, a flickering icon at a given frequency, which through electrical transduction in the eye and transmission by the optic nerves, stimulates an electrical response in the brain at the same frequency. In Chapter 3, the physiological basis and measurement setup of steady-state visually evoked potentials are initially explained. Then, the metrological problem for correcting these elusive signals for the inexperienced technician arises because of their levels on the order of microvolts, and the decidedly unfavorable signal-to-noise ratio. In Chapter 4, techniques for the design, prototyping, and performance analysis of EEG-based wearable systems of steady-state visually evoked potentials are explained. In Chapter 5, three case studies of applications of these responsive BCI systems to industrial maintenance in 4.0 and civil engineering and to the rehabilitation of children with Attention Deficit Hyperactivity Disorder (ADHD) are then illustrated.

The second direction concerns *passive* BCI, in which the brain electric field is measured for the purpose of assessing states of mind, such as stress, attention, engagement, etc. Chapter 6 first focuses on the measurands, which have quantitative roots in psychometrics, the psychological experimental science for the quantitative assessment of human behaviors. The field of study involves two main aspects of neuroengineering research: the development and refinement of metrological foundations of measurement and the construction of procedures and instruments and for wearable measurement. Last but not least, the quality of measurement is also explored here. Chapters 7 and 8 are devoted to the design and prototyping of measurement systems and five case studies, respectively, related to measures of stress, emotional states, attention and engagement. In this exciting and unconventional field of research, in which the possibility of measuring emotional states is going to be made usable at the mass level, the ethical and legal aspects are crucial.

The third direction concerns *active* BCI and specifically the effort to make feasible an ancestral human dream: to move an object with thought. This is technically possible by motor imagery by involving the measurement of cerebral electric field related to the imagination of moving, for example, a limb. To this aim, Chapter 9 deals with fundamentals of motor imagery, either concerning the metrological issues and the neurological phenomena to detect. Then, Chapter 10 presents the design, the prototyping, and the performance assessment of a system based on motor imagery measurement. Finally, two case studies are discussed in Chapter 11, namely the possibility to realize a wearable system for control application by minimizing the number of EEG channels, and motor rehabilitation by means of human-machine adaptation.

Introduction

Brain-Computer Interfaces (BCIs) are systems enabling non-muscular communication and control. Such systems rely on the measurement of brain activity, from signals acquisition to processing, thus aiming to associate a meaning to voluntarily or involuntarily modulated brain waves. A main application of these interfaces is the replacement of lost function for people with severe impairments. In addition, they can also help in restoring, improving, enhancing, or supplementing the nervous system functions. Hence, BCIs can also be employed as assistive devices for people with no impairment. Typical examples are gaming, entertainment, education, or novel tools for workers. Whether these kinds of neural interfaces are sought for able-bodied people or impaired ones, the need to move from laboratory environments to daily-life has recently emerged.

The quest for daily-life BCIS has thus led to new challenges and requirements. Surely, the first requirement is to adopt *non-invasive* neuroimaging techniques to acquire brain activity. This is trivially justified by the need to avoid surgical risks and because of the poor social acceptance of an invasive implant inside the skull. Nonetheless, from a strictly metrological point of view, a relevant drawback is the unavoidable degradation of the signal quality if compared to invasive neuroimaging techniques. Such a degradation is even more present when *wearability* and *portability* constraints are considered. For instance, these typically imply that the electroencephalography should be preferred to the magnetoencephalography, although the magnetic field would not be affected by the presence of the skull as it happens for the electric field associated with brain activity. Furthermore, wearability is correlated with the usage of few acquisition channels, thus requiring to extract the neurophysiological phenomena of interest by less available information. On the other hand, portability implies that motion artifacts could greatly affect the acquired signals, and again this requires a proper processing to remove such artifacts without diminishing the amount of available information about brain activity. Lastly, it is worth mentioning that a user-friendly interface for daily-life applications requires an *ergonomic* acquisition system and minimal calibration time for the system to work. Overall, the challenges associated with this technological trend are related either to the hardware and software part constituting a BCI.

On the road to daily-life BCIs, the illustration of three different paradigms is carried out in this book. Firstly, a reactive paradigm was investigated, namely a BCI relying on external stimulation. Steady-state visually evoked potentials were particularly considered as they have been largely investigated in the scientific community with respect to more classical architectures. Next, passive paradigms were considered, namely BCIs relying on the measurement of involuntarily (spontaneously) modulated brain activity. They are indeed very interesting for monitoring the mental state of a user. In particular, monitoring of emotions, attention, engagement, and stress are pressing issues in different application domains, both in terms of the production process and product innovation. The ongoing technological transformation introduces new opportunity and risk specifically connected to the new framework of very high human-machine interaction. Passive BCI is a promising channel to improve the adaptivity of a cyber-physical system to human.

Finally, an active paradigm was considered, whose operation relies on spontaneous and voluntarily modulated brain activity. Notably, motor imagery was exploited, i.e., the act of imagining a movement without actually executing it. Although more interesting because of the possibility to avoid external stimulation, the detection of motor imagery is usually more critical than detecting evoked potentials.

The book has been structured accordingly. Chapter 1 gives an essential background on BCI technology. Its history is briefly retraced from electroencephalography invention to brain-races, then these systems are categorized according to their different characteristics, and finally a general architecture is presented by highlighting the main blocks. The chapter ends by extensively discussing the requirements and reporting some implementation examples pointing to daily-life neural interfaces. Moreover, some perspectives are outlined as possible developments addressed to a near future. Chapter 2 focuses on the design of daily-life interfaces that, with an abuse of notation, are referred to as "wearable". This keyword stresses the need to use little cumbersome hardware for both signal acquisition and actuation, and this is even related to low-cost and user-friendly equipment. After these general chapters, the discussion of the following chapters is devoted to the specific realization of wearable interfaces in the reactive, passive, and active contexts, based on visually evoked potentials, mental state detection, and motor imagery, respectively. In particular, Chapters 3, 6, and 9 report some considerations on the phenomena definitions, on its neurophysiological basis and state of the art of each paradigm analyzed. Chapters 4, 7, and 10 present the design, prototyping, and experimental validation of such systems. Finally, applications are even discussed in Chapters 5, 8, and 11 in order to better justify the relevance of the illustrated systems.

Part I

Background

1

Electroencephalography-Based Brain-Computer Interfaces

The first chapter of this book provides an overview on Brain-Computer Interfaces (BCIs) based on electroencephalographic signals. In particular, the history of this technology is briefly retraced in Section 1.1, a practical taxonomy is reported in Section 1.2, while the description of a general BCI architecture is presented in Section 1.3. Next, in Section 1.4 the main measurement approaches of brain phenomena are introduced and a focus on electroencephalographic activity is then given in Section 1.5. Finally, the reactive, passive, and active BCI paradigms are described in Section 1.6.

1.1 HISTORY

In 1924, the German psychiatrist Hans Berger (1873 – 1941) was able to record for the first time the electrical activity of the human brain. After 1924, neurophysiologists continued to record a big amount of brain signals in several conditions. In 1929, Berger published a report with data obtained from 76 subjects of a research study carried out by introducing electrodes in their skull [1]. Moreover, some researchers focused on voluntarily control of the brain rhythms [2, 3]. The expression *Brain-Computer Interface* (*BCI*) was first introduced in 1973, in the famous article by Jacques J. Vidal *Toward Direct Brain-Computer Communication* [4]. In particular, his research at the University of California demonstrated the feasibility of a communication relying on the information extracted from electrical brain waves. In a further article, *Real-Time Detection of Brain Events in EEG* [5], Vidal showed that, in some situations, it was possible to detect and reliably classify in real-time evoked responses or event-related potentials. The system he proposed required a stimuli generator for the elicitation of brain potentials. In particular, the user was able to drive a mobile cursor through a maze by focusing intermittent stimuli on a display. The system recognized the direction desired by the user thanks to the brain signal patterns stimulated by the different focused stimuli.

Due to the increasing number of research groups dealing with the necessity of establishing common definitions, methods, and procedures on BCI

DOI: 10.1201/9781003263876-1

(more than 20, at that time), the need to organize meetings arose. The first international meeting on BCI research was organized in 1999, more than 25 years later, and a review of this event was published on the *Institute of Electrical and Electronic Engineers* (*IEEE*) journal Transactions on Rehabilitation Engineering [6]. Essentially, all BCI systems were based on *Electroencephalography* (*EEG*), and their performance was assessed with a parameter called *Information Transfer Rate* (*ITR*), i.e., the amount of information transmitted per unit time, whose maximum value was reported to be 5–25 bit/min. It was also highlighted that the central element in each BCI system is the *algorithm translating electrophysiological input from the user into output that controls external devices.* These considerations were mostly confirmed in the second international meeting, held in 2002 [7]. After 30 years from the first proposal of a BCI, the results were already poor, although advancing technology and more powerful computers were unlocking new horizons. The achievement of higher speeds and greater classification accuracies, both implying greater ITRs, led back to the need for improving signal acquisition and processing, but also to a better user training. It was clear that such improvements depended (and depend today) on a synergy between different disciplines and on the adoption of common criteria and strategies among the ever-growing research groups.

In the last decade, BCI technologies have seen considerable acceleration in their development, mainly because of the advances in processing algorithms [8], but also because of the progress in neuroimaging techniques [9]. Some BCI paradigms are currently reaching 100 bit/min, though there is a trade-off between such a performance and easiness of use [10]. For instance, high ITRs are typically reached with a BCI relying on external stimulation, and this might not always be desirable. Furthermore, the ITR is not necessarily an appropriate metric to describe the performance of a BCI system, and the definition of ITR itself has been questioned in some contexts [11]. In details, ITR calculation basically assumes that the BCI system has no memory effect, and the ITR could be badly estimated when the number of available trials is low (see [11] for further details and deeper discussion).

As mentioned above, the BCI research has always been related to medical applications, promising solutions for people affected by amyotrophic lateral sclerosis or by the locked-in syndrome [12], as well as new rehabilitation possibilities [13, 14], e.g., in post-stroke recovery. At the same time, BCIs can be used by able-bodied people in leisure activities, such as gaming and entertainment [15], or in other daily-life applications like smartphone control [16]. As technological development goes on, BCI communities are growing over the world, and many researchers involved in the field belong to the "BCI society" founded in 2013. Several scientific disciplines cooperate in research and development, and among them it is worth mentioning neuroscience, psychology, engineering, and computer science. In addition, ethics is another important concern that is accompanying the field. Thanks to conferences and other events, the BCI culture is spreading around the scientific and non-scientific communities. Notably the BCI international meeting is currently held every two years, and the Graz BCI conferences are also held every two years in an alternative manner. It is also worth mentioning that "BCI competitions" were held in the recent past (2000–2008) as an occasion for different research groups to compete and interact, and today these events have still an echo because the datasets originated by them are often considered as benchmarks in the BCI literature [17]. Finally, starting from 2016, a BCI-related contest is held as one of the disciplines of the Cybathlon, an international multi-sport event in which people with disabilities compete by exploiting assistive technologies [18].

1.2 TAXONOMY

It is not easy to establish well-defined categories in sorting out systems or technologies, especially for a field like "Brain-Computer Interface" that is complex and continuously evolving. Most probably, such well-defined limits between different things do not even exist. However, the attempt to set those boundaries is an essential task for a better comprehension of the topic. Taxonomy allows to highlight the major differences between various systems and technologies, and this leads to focus on the better solution for a specific problem.

BCI systems can be divided into categories according to several aspects: (i) invasiveness (*invasive/non-invasive/partially invasive*); (ii) the type of input signals (*endogenous/exogenous*) ; (iii) dependence from other brain outputs (*dependent/independent*); and, finally, (iv) timing of interaction between the user and the system (*synchronous/asynchronous*). The first distinction concerns the invasiveness of the technique employed to acquire brain activity. When sensing elements are placed outside the skull, the BCI is categorized as *non-invasive* (Fig. 1.1) This guarantees an easy and safe utilization, but detection reliability is affected by the tissues interposed between the actual signal source and the acquisition system. On the other hand, *invasive* techniques (Fig 1.2) may produce highly reliable control signals with high-speed information transfer. However, there are risks related to the need of surgery for implanting the sensors inside the skull, they are difficult to use, and they potentially lead to infection and long-term viability [19]. These invasive BCIs are often distinguished from *partially invasive* BCIs (Fig 1.3). The difference is that neural probes are not placed inside the gray matter of the brain, but outside the gray matter though inside the skull, e.g., on the cortical surface. It is important to highlight that the difference between them does not merely reside in the level of risk for the user, but also in different measurement reliability. Indeed, long-term signal variability in invasive BCIs is induced as a result of neuronal cell death, cicatrization, and increased tissue resistance [20]. For these reasons, a partially invasive technique like Electrocorticogram (ECoG) reduces the risks and instabilities of an intracortical electrogram while preserving the advantages of a more invasive approach with respect to a non-invasive EEG, such as greater spatial resolution, temporal resolution, and immunity to noise.

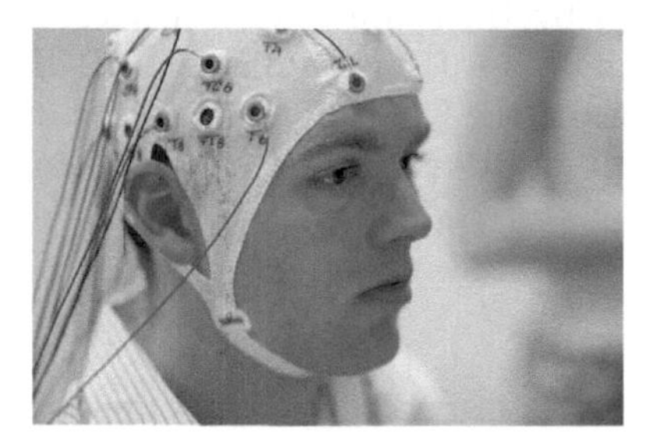

FIGURE 1.1 A classic non-invasive technique for measuring brain activity, the EEG.

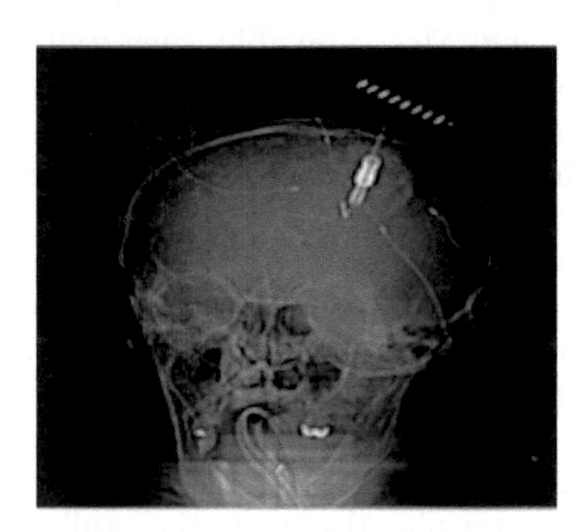

FIGURE 1.2 Representation of an invasive intracortical sensing [21].

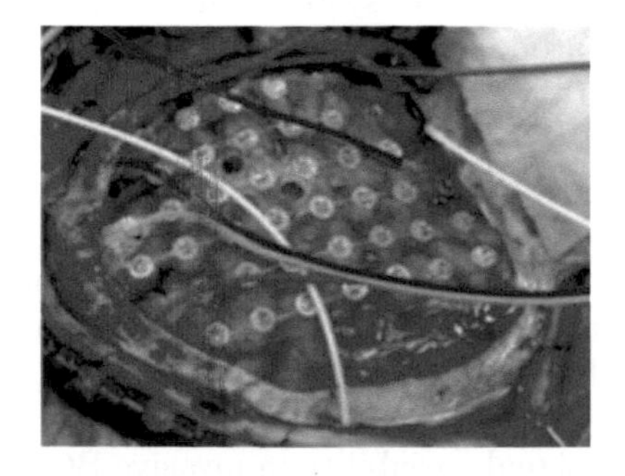

FIGURE 1.3 ECoG as an example of partially invasive measurement [20].

BCIs can also be distinguished according to the nature of the signals they use as input [6]. Notably, *endogenous* control modules respond to spontaneous control signals from the user, e.g., a motor imagery task generating Sensorimotor rhythms (SMRs), while *exogenous* control modules respond to control signals evoked from the user by a stimulus [20]. At this level, the distinction is thus between stimulus-related paradigms (exogenous) versus non-stimulus-related (endogenous). If a stimulus is needed, a further distinction can be done among evoked and induced potential. *Evoked potentials* are oscillations that are phase locked to the stimulus, and they can be detected by analyzing the average among different trials or with a spectral analysis. On the other side, *induced potentials* are not phase locked to the stimulus, and they are detected by first removing the average component among trials and then analyzing trials in time-frequency [22].

Another distinction is made between dependent and independent BCIs [23]. A *dependent* BCI relies on other normal brain outputs, usually a muscular activity. An example is a BCI based on the *Steady-State Visually-Evoked Potential* (*SSVEP*). This evoked potential is generated in the occipital area of the brain when the user stares at a flickering icon. This usually requires that the user moves his/her eyes to gaze at a specific icon. On the other side, an *independent* BCIs provide a completely new communication channel because it does not depend on already existing pathways. To make things more confused, a study suggested that also SSVEP can be used without a gaze shift, and in that case such a system would be considered as an independent BCI [24]. In any case, this distinction helps to identify applications that can still be used in the event of total loss of muscle control (e.g., advanced stage Amyotrophic Lateral Sclerosis).

Finally, a last functional distinction is worthily mentioned. A BCI system is *synchronous* if the interaction between the user and the system occurs over a specific period of time, which has also a well-defined "cue"). Therefore, only in that period the device can receive biomedical signals and process them. Instead, if the user wants to interact at any time, there is the need of an *asynchronous* BCI, which is able to react to the mental activities coming from the patient at any time without any restrictions. The last aspect is quite important in daily-life applications, but the asynchronous mode clearly introduces more difficulties because the system must be not only capable of discriminating "a what" but also "a when" [25].

A further distinction was proposed in [26] between *reactive, active, and passive* BCIs. In a *reactive BCI*, the brainwaves are produced in response to external stimuli, in a *passive BCI* the user does not consciously control his/her electrical brainwaves, while, in an *active BCI* the modulation of the brain waves is voluntary, in order to control an application, independently of external events. This distinction is explored in more detail in Section 1.6 and it is essential to fully understanding the ensuing discussion.

1.3 ARCHITECTURE

A general-purpose BCI architecture is represented in Fig. 1.4. Signals are acquired from the user's scalp, and then processed to retrieve a command for the application of interest. The signal processing can be divided in two stages: (i) features extraction, for deriving a meaningful synthetic description of the signals, and (ii) features translation, for interpreting them by giving back an output command. A feedback is finally given to the user depending on the application.

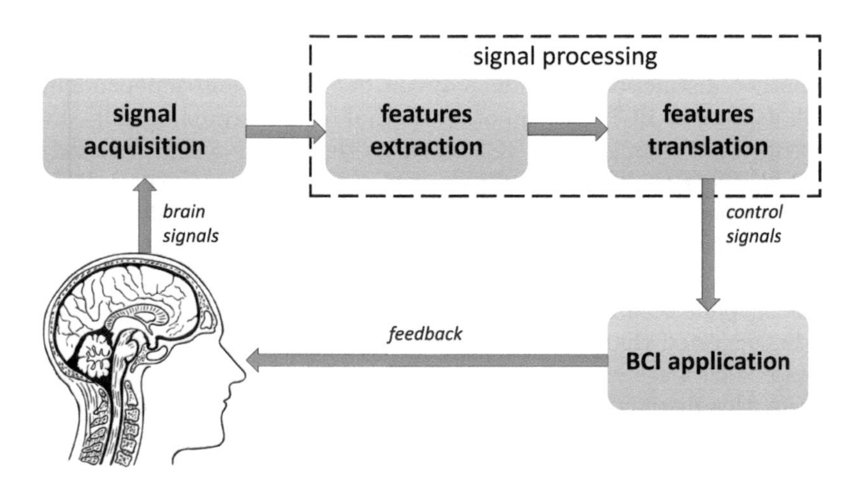

FIGURE 1.4 General architecture of a brain-computer interface.

In the following, the blocks of the BCI architecture are detailed. The purpose is to give a deeper understanding for the different elements constituting such systems, with a particular emphasis on EEG-based systems. Meanwhile, further details are addressed to the specific implementations that are treated in the next chapters.

1.3.1 Signal Acquisition

Different neuroimaging techniques, i.e., techniques to map the structure and function of the central nervous system, can be used in BCI systems [27]. The electromagnetic activity is typically measured, but there are also other physiological phenomena related to brain activity, e.g., blood dynamics. The interested reader will find in Section 1.4 an extensive discussion of the major neuroimaging techniques. Concisely, the brain activity is described by means of signals in time domain related to different neurophysiological phenomena. The probes for acquiring these brain signals can be put inside or outside the user's scalp, while determining the invasiveness level of the BCI system. In the present book, the focus is on electrodes placed outside the scalp to acquire the cortical electrical activity. Therefore, a non-invasive electroencephalography (EEG) will be considered. Despite the enormous advantages of EEG non-invasiveness in practical applications, its spatial resolution is limited (order of magnitude: 1 cm). In fact, the signals measurable from the scalp surface can only be related to populations of active neurons [28]. This is indeed one of the major factors limiting neuronal activity decoding in EEG-based systems. Furthermore, the electrical activity is attenuated by the layers between the brain and the electrodes, thus it is more sensible to noise.

Electrode types can be distinguished between *wet* and *dry*, depending on whether conductive gels are used or not at the electrode-scalp contact. Gels used in wet electrodes have the advantage to enhance the contact with the skin and, hence, improve the signal quality. They are used in conventional systems, notably in clinical application. On the other side, dry electrodes guarantee higher comfort for the user and they are more suitable for daily-life applications. A recent comparison between these two technologies showed that signal quality complies with the needs of clinical applications even when dry electrodes are used [29]. However, particular attention must be made since these electrodes have a poorly stable contact and hence they are more affected by artifacts during recording. For both the electrode types, pre-amplification or simply buffering can be considered for impedance matching: such electrodes are thus referred to as *active*. In contrast, *passive electrodes* (no pre-amplification) are simply conductor and they are usually employed for signal referencing.

Electrodes placing was standardized with the 10-20 system [30]. This measurement technique is based on standard landmarks of the skull, namely the nasion, which is a craniometric point placed in the midline bony depression between the eyes where the frontal and two nasal bones meet, the inion, other craniometric point that is an external occipital protuberance, and the left and right pre-auricular points. The pre-auricular points are felt as depressions at the root of the zygoma, which are behind the tragus [30]. The system's name is derived from the fact that electrodes are placed in determined positions at 10 % intervals and 20 % of the distances joining the nasion-inion points and the right and left pre-auricular points. The standard electrode locations of the 10-20 system are shown in Fig. 1.5, where the positions *Fpz* and *Oz* are highlighted as an example. These specific positions will be considered in Chapter 3 for the measurement of visually evoked potentials. It is also worth mentioning that more electrode locations can be identified with the 10-10 system, which defines a finer spatial grid for the placement.

With particular regards to EEG acquisition, signal amplitudes normally span the 0.5 µV to 100 µV range [28]. Therefore, amplification is another crucial part of the acquisition system. Together with filtering, it contributes to the signals conditioning needed before voltage acquisition. Frequency bands of interest are typically limited to about 100 Hz, but this limit can be lowered depending on the actual case. Clearly, filtering should always consider

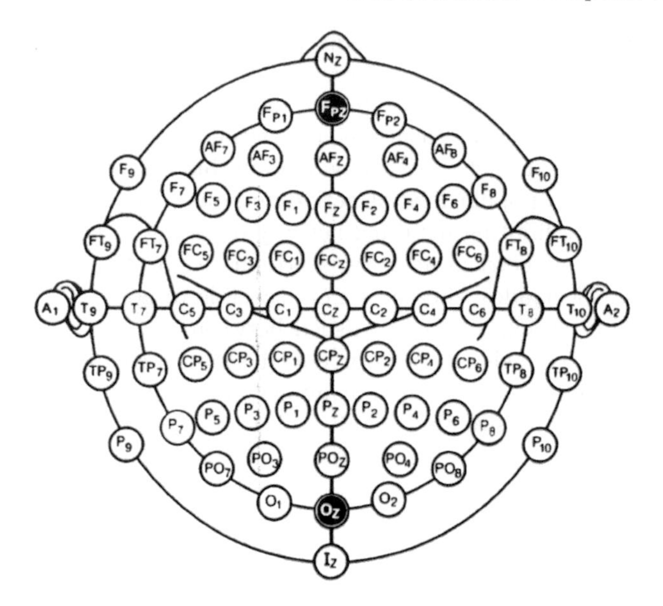

FIGURE 1.5 EEG electrodes locations according to the international standard framework 10-20 [30]; the positions *Fpz* and *Oz* are highlighted in black at the scalp frontal region and at the occipital region, respectively [31].

the only frequencies of interest in trying to maximize the signal-to-noise ratio, which is quite limited in EEG measurements. Moreover, the advent of digital electronics, therefore the voltage acquisition consists of an Analog-to-Digital Conversion (ADC), so that the digitized signal can be better processed by the following stages and stored in digital memories. In few cases, analog EEG acquisitions are still taken into account. They consist of voltage-controlled deflections of pens writing on a paper tape. Such analog EEG machines are certainly a historical heritage still popular among clinicians, but they are definitely not used in an engineering context. In either case, the vast majority of EEG acquisition systems consider multi-channel devices with 64 or more electrodes. Acquisitions are often triggered in order to synchronize measurand signals with an external event. This is typically the case when event-related potentials are measured, while, generally speaking, there is the need to provide timing of data points in most BCI applications. More details on the measurand signals are recalled in the Section 1.5 with respect to classical frequency bands and common paradigms. Nonetheless, other relevant aspects will be better treated in the following chapters by proposing specific BCI systems design.

1.3.2 Features Extraction

Once the signals are acquired, the decoding of neural activity requires proper processing. In the first processing step, peculiar features must be extracted from available signals in order to synthetically describe the signals while trying to emphasize the informative content. This is useful for the interpretation of the user intention or mental state. An example of features extraction is represented in Fig 1.6. When the signal to analyze is represented in frequency domain (Fig. 1.6a), the plot shows some harmonic components that could characterize the signal. However, in aiming to distinguish this signal from another one, a suitable choice could be describing the signal in terms of the power associated with the harmonics. Assuming that these signals can be described with two power features, each one corresponds to a dot in a plane. Such a *features domain* representation can highlight a separability between two classes of signals (Fig. 1.6b).

Digitized EEG signals are affected by noise, either electrical or biological interferences, therefore pre-processing is often considered before the features extraction. This phase usually consists of spatial and/or frequency filtering and decomposition, so to allow artifacts removal, or at least their reduction. They mostly require multi-channel acquisitions, or even additional signals coming from electrooculography and/or electromyography [32]. For instance, eyes and muscular movements affecting the EEG signals can be identified with a proper decomposition in order to remove the artifacts-related components. Indeed, in doing that, there is also the risk of removing relevant information. After the pre-processing, commonly used features in representing EEG signals are *frequency band power features* and *time point features* [8]. Band powers represent the signals for a fixed frequency band and a fixed channel with respect to its mean power or mean energy over time. The time windows considered for this calculation are referred to as *epochs*. On the other hand, time point features concatenate samples from available channels. Band power features are mostly used when the amplitude of EEG rhythms are related to the neurophysiological phenomenon of interest, while time point features aim at describing temporal variations. Such considerations suggest knowledge-driven processing approaches. Notably, exploiting the knowledge about neurophysiological phenomena is desirable when there is a limited amount of available data to train a model for feature extraction. This prevents the risk of overfitting the model on training data, which would lead to poor performance on independent data. For instance, a deep neural network could be more prone to overfitting since it typically involves many parameters to be determined, while available EEG data could be not sufficient for that. In this context, an interesting research trend is also attempting to explain *Artificial Intelligence (AI)*, so to validate the training of the features extraction models while providing an interpretation of the measurand signals features. This appears particularly relevant in medical and biomedical applications.

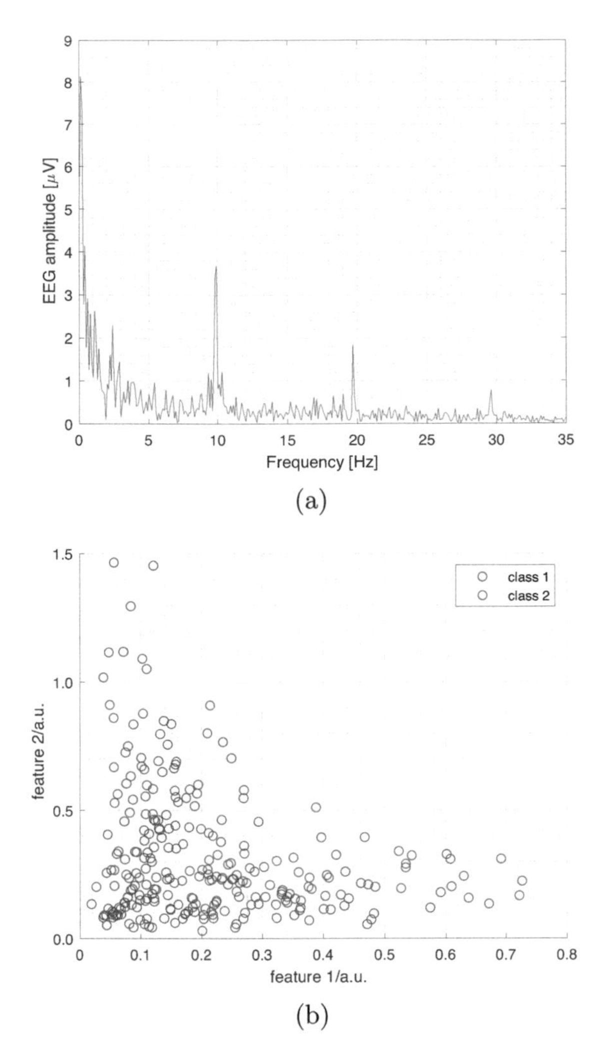

FIGURE 1.6 Feature extraction for maximizing separability of signals from two different classes: representation in frequency (a) and feature (b) domains (a.u.: arbitrary unit).

Another crucial step is features selection. This step allows to reduce the number of predictors, which represent the signals under analysis, to a subset of mostly informative features. The aim is again to reduce the risk of overfitting the signal processing model on training data by excluding non-relevant features, thus enhancing the predictive accuracy and the interpretability of the model. Three main selection methods include *subset selection*, *shrinkage*, and *dimension reduction* [33]. *Subset selection* can consider all possible subsets of features to find an optimal one by relying on an objective function. However, computational burden is often prohibitive, therefore stepwise selection is typically considered, since it allows to add or remove a single best feature at time. In *shrinkage* methods, the selection is conducted by reducing coefficients associated with each feature toward zero: this method inevitably involves the features translation step, in which a functional relationship between features and model response is attempted. Lastly, in *dimension reduction* methods, the directions of maximum variance in the features space are typically derived, e.g., with Principal Components Analysis. It is worth noting that the feature selection step is often integrated within the feature extraction itself, or it can be carried out in conjunction with the training of the features translation model (regression or classification). Broadly speaking, there could be no clear distinction between the blocks of signal processing: the current discussion considers separated blocks for simplicity, but features extraction, selection, and translation can be conducted simultaneously, such as in deep neural networks.

1.3.3 Features Translation

In the general BCI architecture discussed here, the last part of signal processing consists of translating the features describing brain signals into control signals for an application. The goal is to derive a functional relationship between the features (predictors) and the output control signals (response), usually by regression or classification. Classification is mostly considered, because, in most cases, there is no logical order among the possible output values. In particular, the measured brain signals can be associated to a specific class between two or more choices.

Modern classification methods mostly rely on Machine Learning. As a first classification attempt, a simple approach is typically considered, such as *Linear Discriminant Analysis* (*LDA*) or *k-Nearest Neighbors* (*kNN*) [33].
Linear Discriminant Analysis [34] is a dimensionality reduction algorithm used to maximize the separability between data points belonging to different classes. It works by projecting the features from higher dimension space into a lower dimension space, while preserving the discriminatory information of the data of different classes.
k-Nearest Neighbors [35]is a non-parametric method that works by assigning the class of its nearest points to an unlabeled data point.

Successively, a more complex approach can be adopted to enhance the classification, such as *Decision Trees*, *Random Forest* (*RF*), *Naive Bayes* (*NB*) classifiers, *Support Vector Machine* (*SVM*) with linear or non-linear kernels, and *Artificial Neural Network* (*ANN*). In general, when the classification approaches rely on the features statistical distribution, one can also speak of *Statistical Learning.*
In particular, *Decision Trees* [36] are non-parametric Supervised Learning algorithms for classification and regression. They work by creating a model that predicts the class of a variable by learning decision rules from the data.

Random Forest [37] is a Supervised Machine Learning Algorithm for classification and regression that works by building a multitude of decision trees on different samples at training time. For classification tasks, the output of the RF is the class selected by most trees.
Naive Bayes [38] are a sets of classification algorithms based on Bayes'theorem. They end up being very useful as a quick-and-dirty baseline for a classification problem because of their speed and the low number of parameters to tuning. All Naive Bayes classifiers assume that the value of a particular feature is independent of the value of any other feature; if this assumption is true, the algorithms' accuracy is comparable to that of artificial neural network and decision trees classifiers.
Support Vector Machine [39] is a binary classifier which classifies input data according to the decision hyperplane having the largest distance from the *margins* of the classes [40].
Finally, *Artificial Neural Networks* [39] are Deep Learning algorithms whose structure is inspired by the human brain. It is composed by a set of *neurons* linked to each other into fully connected layers. The output of each neuron is a linear combination of the inputs followed by a non-linear function, which is referred to as *activation function.*

Figures 1.7 show two different datasets represented in bi-dimensional features domains to highlight that the most suitable classification approach depends on the shape of the boundary between different classes.

Furthermore, in the last years, novel approaches have been developed starting from the abovementioned classical methods [8]. Notably, adaptive classifiers are used to deal with non-stationarity signal, tensor-based classifiers are used to analyze data in a geometrical space allowing better manipulation, or Transfer Learning techniques are explored in order to generalize a trained model to data with a different statistical distribution. Deep Learning has also attracted many researchers in the BCI field, but it will not be considered hereafter since it typically requires more data.

Classification performances can be assessed by different metrics depending on the final application. A largely considered parameter is the *classification accuracy*, which is calculated as the ratio between correctly classified signals divided by the total number of classified ones. This accuracy is actually a success rate, and it quantifies how many times the BCI system correctly recognizes the user's intention. Classification accuracy is also a common objective function in training the features selection part of the model. Its calculation requires that test data, independent of the training data set, are available. However, during the training procedure, classification accuracy can be estimated with *Cross Validation* (*CV*). This procedure consists of randomly splitting training data into training and validation subsets, so to estimate the classification accuracy by averaging the accuracies on the validation subsets across different iterations. CV is extensively used in model selection during training and for estimating the model performance on independent data [41].

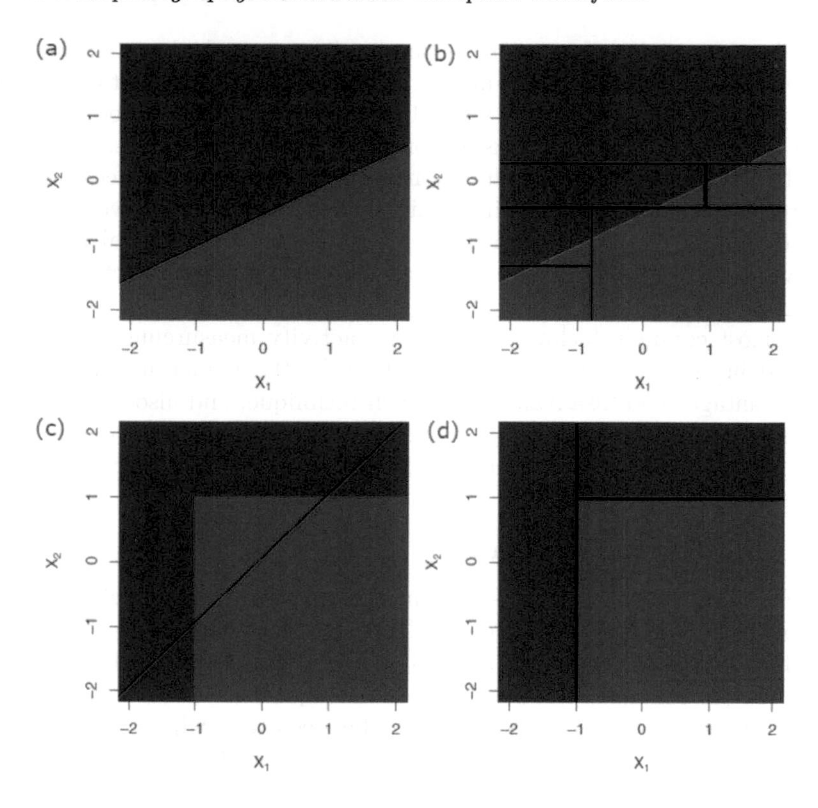

FIGURE 1.7 Linear and non-linear classification of two different datasets represented with two features. The classifier boundaries are in black, the two possible classes are in red and blue: (a) linear classifier used for linearly separable classes, (b) non-linear classifiers attempting to separate linearly separable classes, (c) linear classifier attempting to separate non-linearly separable classes, and (d) non-linear classifier used for non-linearly separable classes [33].

1.4 NON-INVASIVE MEASUREMENT OF NEURAL PHENOMENA

The core of a BCI system is the acquisition and processing of brain signals. The choice of a proper approach for signal processing largely depends on the brain signals to be analyze, and most of BCI literature has been focusing on algorithms of varying complexity for the analysis of specific brain activities [8]. Hence, it is of foremost importance to select a proper signal acquisition method in accordance with the requirements. In line with our current understanding of the human brain, the measurand brain activity is related to neurons, hundreds of billion nerve cells communicating by electrical signals or chemical reactions.

Typically, the activity of an entire brain area is measured, thus meaning that the average activity of a great number of neurons is considered at once. Each area is specialized for particular tasks, but concurrent tasks can be processed by the same area and many tasks are processed in multiple areas. Advancing technology and more deep brain function understanding allow to decipher with ever greater resolution the ongoing brain activity, but it is not (yet) possible to "read thoughts" [27]. Nowadays, one can only measure general processes, and several challenges are posed by intra-subject and inter-subject variability in brain functionality.

The most common techniques for brain activity measurement are briefly described in the following. The aim is to help the reader understand the main advantages and disadvantages of each technique, and also to justify the focus on electroencephalography. Interestingly, multiple measurement techniques could also be used at once in order to improve BCI performance [19].

1.4.1 Electroencephalography

Electroencephalography (EEG) is a technique used to record the electrical activity of the human brain by means of electrodes placed on the scalp [28]. The term "EEG" typically refers to a non-invasive technique. The measurement setup is simple and safe. However, the electrical potentials detected with this technique must cross the tissues interposed between the scalp and the brain to reach the electrodes. Consequently, there is a considerable attenuation and the detection reliability decreases. Instead, in an ECoG, the electrodes would be surgically positioned on the cerebral cortex, or they could be even implanted inside the cortex in intracortical techniques or electrograms. This would guarantee greater temporal and spatial resolution, as well as greater signal-to-noise ratios if compared to an EEG [20]. Usually, an EEG acquisition system is inexpensive, highly portable, and wearable. Clearly, this also depends on the number of electrodes, which can be more than one hundred. Typically, electroencephalography guarantees a temporal resolution of 10 ms and a spatial resolution of 1 cm [27, 42]. Electrical signal amplitudes vary from 5 µV to 100 µV for EEG, while these signal amplitudes would reach 1 mV to 2 mV if the electrodes were implanted in the skull [43]. The signal bandwidth of usual interest spans in the range 0.1 Hz to 100.0 Hz, and the amplitude decreases at increasing the frequency. Recent studies also support the idea that EEG amplitudes correlate with the degree of neuron synchronization [44]. On the other side, EEG alone cannot detect brain area activation occurring in deeper regions. In this regard, techniques like Functional Magnetic Resonance Imaging (i.e., a technique exploiting nuclear magnetic resonance to reconstruct an image), and magnetoencephalography (i.e., a technique allowing the detection of magnetic fields generated by the neurons' electrical activity) can be employed.

1.4.2 Magnetoencephalography

Magnetoencephalography (*MEG*) is a technique allowing the detection of magnetic fields generated by the neurons' electrical activity. This technique guarantees deeper imaging if compared to the EEG, with a greater spatial resolution (about 1 mm) and a similar temporal resolution (1 ms to 10 ms) because the skull is almost completely transparent to magnetic fields [27, 42]. However, it is an expensive and not portable technology owing to the need of big magnets, and/or superconductivity, e.g., in superconducting quantum interference device [20]. Furthermore, MEG equipment can interfere with other instrumentation or suffer from electromagnetic interference. Therefore, MEG seems not very suitable for applications in every-day life. Nonetheless, MEG is seen as complementary to EEG: a synergically employment of both techniques can provide important information about the dynamics of brain functions [45].

1.4.3 Functional Magnetic Resonance Imaging

Magnetic Resonance Imaging (*MRI*) exploits nuclear magnetic resonance to reconstruct an image of a physiological process. A strong magnetic field is generated to make atoms absorb and release energy at radio frequency. In particular, hydrogen atoms are usually involved, which are abundant in water and fat. By localizing the signals emitted from the atoms, it is thus possible to obtain an image [46]. There are also other imaging techniques such as *Computed Tomography* (*CT*) or *Positron Emission Tomography* (*PET*), but despite them MRI does not use X-rays or ionizing radiation.

Functional Magnetic Resonance Imaging (*fMRI*) exploits the basic principles of MRI to detect changes in blood flow through changing blood magnetic properties, which are related to neural activity in the brain. Indeed, neurons need for more oxygen and glucose when they communicate, thus causing an increase in blood flow to active regions of the brain [27]. fMRI is expansive and not portable because it requires cumbersome instrumentation, such as superconductive magnets [27, 42]. It provides data with high spatial resolution (even below 1 mm) but with a poor temporal resolution (few seconds) [42] due to the slow changes in blood flow. A hybrid system could, for instance, employ fMRI in conjunction with EEG or MEG to achieve both high temporal and spatial resolution [45]. Obviously, this would require a proper fusion of the two measurement methods to combine data and provide insight not achievable by a single technique.

1.4.4 Functional Near-Infrared Spectroscopy

Functional Near-Infrared Spectroscopy (*fNIRS*) exploits light in the near-infrared range to determine the blood flow related to neuronal activity. It provides high spatial resolution (few millimeters), but poor temporal resolution (few seconds) [20, 42]. Compared to fMRI, fNIRS is portable and less expensive, but it provides less imaging capabilities. The potential advantages of this technique include insensitivity to signal perception errors (artifacts) typically present in EEG [47]. Hence, a hybrid BCI would benefit from a proper combination of EEG and fNIRS.

1.4.5 Other Techniques

In this subsection, few notes are reported about further measurement techniques related to brain activity detection beyond the aforementioned non-invasive techniques. Neuroimaging techniques like ECoG or intracortical neuron recording are of great interest in the BCI because they allow for a superior signal quality. ECoG is usually referred to as *partially-invasive* or *semi-invasive* to highlight its lower degree of invasiveness if compared to intracortical recording. However, the ECoG itself requires a craniectomy by which the electrodes are directly placed on the exposed surface of the brain [48]. Meanwhile, intracortical neuron recording exploits microelectrode arrays placed inside the gray matter of the brain in order to capture spike signals and local field potentials from neurons [49]. Invasive techniques also guarantee lower vulnerability to artifacts such as blinks and eye movement. Nevertheless, despite the advantages from the metrological point of view, there are evident difficulties in considering such techniques in daily-life applications. Even the already mentioned CT and PET are noteworthy because they are of utmost importance for brain activity monitoring, notably in clinical applications.

CT is a technique that has much in common with PET, except that it uses X-rays and a camera to detect and record the activity of the brain. A similar technique is the *Single-Photon Emission Computed Tomography* (*SPECT*) which instead uses γ-rays [27]. Instead, PET is a nuclear imaging technique based on the detection of blood flow and glucose consumption through a small amount of radioactive isotopes introduced into the blood. This technique has high spatial resolutions, down to 2 mm, while temporal resolution is again limited by the dynamics of [45]. In a certain sense, also CT and PET are invasive and clearly unsuitable for a BCI. Indeed, they are poorly considered in the BCI community, unless they would be needed for a medical diagnosis.

In conclusion, Tab. 1.1 resumes the main characteristics of the described neuroimaging techniques. A particular focus is given to the order of magnitudes for the respective temporal and spatial resolutions, and also to the advantages that make a technique suitable for daily-life applications, as well as the disadvantages that eventually make it unsuitable for that purpose.

As a final note, there are two other measurement techniques that are noteworthy for BCI application, but strictly speaking, they do not measure brain activity. These are the *Electroculography* (*EOG*), which measures the electric potential between the front and the back of the human eye, and the *Electromyography* (*EMG*), i.e., the measuring of the electrical activity of muscle tissue. EOG and EMG artifacts are indeed the most important sources

TABLE 1.1 Summary of the main characteristics concerning neuroimaging methods.

Neuroimaging technique	Physical property	Temporal resolution	Spatial Resolution	Advantages	Disadvantages
EEG	electrical potential	10–100 ms	1 cm	low cost wearabile portabile	noisy sensible to artifacts electrodes placing
MEG	magnetic potential	10 ms	1 mm	deep imaging	expensive bulky
fMRI	blood flow	1 s	1 mm	deep imaging	expensive bulky
fNIRS	blood flow in cortical tissue	1 s	<1 cm	low cost wearable portable	no deep imaging low time resolution
ECoG	electrical potential	1 ms	1 mm	signal quality portable	(semi)invasive
Intracorical	local electrical field potential	1 ms	<1 mm	signal quality portable	invasive
SPECT	blood flow	1 s	1 mm	deep imaging	expensive bulky radiation
PET	blood flow	1 s	1 mm	deep imaging	expensive bulky radiation

of errors in neuroimaging, thus the measurement of ocular and/or muscular activity is exploited in many artifact removal techniques [32].

1.5 MEASURING THE ELECTRICAL BRAIN ACTIVITY

As mentioned above, electrodes used for electroencephalography can be wet or dry. For wet electrodes, gel is needed between the electrode and the scalp to reduce the contact impedance down to 1 kΩ to 10 kΩ. However, the main drawback of gels or electrolytic paste is the long preparation time of the subject and periodic refresh required for good quality signal [49]. Indeed, gel progressively dries out: this determines contact impedance and signal quality to be negatively affected. Moreover, care must be taken to ensure that the gel does not slip between the electrodes since this would create a short circuit. On the other side, dry electrodes do not require any gel. These kinds of electrodes may be dry active electrodes, which have pre-amplification circuits for dealing with very high electrode-skin contact impedances, or dry passive electrodes, which have no active circuits, but are linked to EEG recording systems with ultra-high input impedance.

Usually, in multi-channel recordings, the 10-20 electrode system [30] is adopted. Preparation time of the subject is reduced by already setting pre-assembled electrodes in standard positions on the headset. The amplitude of oscillations at the brain surface can reach tens of mV, but amplitudes recorded on the scalp are about hundreds of µV [50]. Then, in EEG signal, frequency of interest lies in the range 0.5 Hz to 100.0 Hz, and specific sub-bands are related to physiological or pathological states [50, 49]. These sub-bands are know as *EEG rhythms* and five are typically distinguished:

- Delta (δ) rhythms: delta waves have frequencies in the range 0.5 Hz to 4.0 Hz, and their amplitude is usually under 100 µV. In adults, delta waves are associated with a state of deep sleep. Delta waves are mostly present in children. Their amplitude decreases with increasing age.

- Theta (θ) rhythms: theta waves have frequencies in the range 4.0 Hz to 7.0 Hz, and their amplitude is below 100 µV. As the rhythm delta, theta waves are mostly present in children. In adults, theta waves are associated with states of sleep or meditation. In some adults,the rhythm theta is also associated with emotional stress.

- Alpha (α) rhythms: alpha waves have frequencies in the range 8.0 Hz to 13.0 Hz, and their amplitude is below 10 µV. These waves can be recorded during waking state, but they indicate a state of relaxation. In the same range of the alpha waves, but in the motor cortex, the rhythm mu (μ) is also detected. This rhythm is interesting because it is strongly related both to movement execution, and to imagination or observation of the movement executed by someone else (because of mirror neurons [51]) [48, 52].

- Beta (β) rhythms: beta waves have frequencies in the range 13.0 Hz to 30.0 Hz, and their amplitude is below 20 µV. These waves appear during waking state, when the subject is occupied in a mental activity [48, 53]. Beta rhythms are also associated with motor activity. They are modulated either during real movement or during motor imagery.

- Gamma (γ) rhythms: gamma rhythms have frequencies over 30 Hz and they indicate a state of deep concentration. Some experiments have revealed a relationship in normal humans between motor activities and gamma waves during maximal muscle contraction. Gamma rhythms are less commonly used in EEG-based BCI systems, because artifacts or electrooculography are likely to affect them.

1.5.1 Measurand Brain Signals

With the different neuroimaging techniques reported in the previous section, several types of brain signals can be measured. Indeed, there is a relation between the brain activity of interest and the neuroimaging method to employ for its detection. In the present section, the main focus will be on the brain signals measurable through the EEG. However, this does not imply that EEG is the only possibility for such measurements, and surely hybrid approaches are often possible. Then, depending on the available signals and on the final application, a proper processing approach will be adopted.

In following a chronological order, it is convenient to start by introducing *Evoked Potentials* (*EPs*). EPs are variations of the EEG signals occurring as a result of a sensory stimulation. For instance, a largely treated class of EPs are the *Visually-Evoked Potentials* (*VEPs*), where brain activity modulations occur in the visual cortex in correspondence of a visual stimulus [54]. Typical visual stimuli are flickering icons or light flashing. Therefore, the VEPs reflect the brain's processing of visual information [55] and different types can be distinguished according to the kind of information they contain. An interesting distinction is proposed in [56] between *transient VEPs (t-VEP)*, *steady-state VEPs (SSVEP or f-VEP)*, and *code-modulated VEPs (c-VEP)*. In t-VEPs, the frequency of visual stimulation is kept below 4 Hz to 6 Hz so that consecutive flashes do not overlap. The target flashes are mutually independent and they are typically detected by averaging over many signal epochs. Hence, successive epochs must be properly synchronized. Meanwhile, in *frequency modulated VEPs (f-VEPs)*, more commonly referred to as steady-state VEPs (SSVEPs), the visual stimulus is a flickering icon with a frequency above 6 Hz. In such a case, the evoked potentials result from the overlapping of consecutive flashes, whom period is lower than a single t-VEP duration [55]. Different targets are therefore distinguished by means of different flickering frequencies and/or thanks to the phase information. Where the phase relation between targets is of interest for the SSVEP detection, a trigger signal is also needed for synchronization. Finally, c-VEPs rely on pseudo-random sequences of bits that modulate the duration of a flashing target. In detecting the response to a code-modulated stimulus, a synchronization of measured signals is needed and a template matching method is exploited to retrieve the code. Higher communication speeds are achievable with a c-VEP [56], but necessarily required user and algorithm training. Instead, a trainingless BCI can be built, at least in principle, by considering SSVEPs. Another important difference between c-VEP and SSVEP is that the first has a broadband spectrum, while the second is characterized by narrow bands. Such considerations are crucial in choosing the processing approach. Generally speaking, EPs recorded with EEG have relatively small amplitudes, so there is the need of proper filtering and amplification to extract the signal features of interest [49].

The discussion conducted for visually EPs can be extended to other evoked potentials, relying for example on auditory or somatosensory stimulation. As a whole, EPs are common measurands in reactive BCI paradigms and, thanks to the EEG, they can be exploited in building non-invasive, wearable and portable BCIs for daily-life applications. In particular, SSVEP-based BCI systems are considered in this book. Meanwhile, it is worth reporting further considerations about a largely exploited transient EP, which is the P300 potential. P300 EPs are positive peaks in the EEG elicited by infrequent visual, auditory, or somatosensory stimuli (flashes) [10]. The P300 wave is an endogenous response elicited about 300 ms after attending the stimulus. As already highlighted for SSVEPs, the use of P300-based BCIs does not necessarily require training. However, some studies proved that the less probable the stimulus, the larger the amplitude of the response peak, and for that reason the system performance may be reduced if the user gets used to the flashes [49]. Also, P300-based BCIs speed is limited by the averaging process that is usually required in transient EP detection. On the other hand, it is possible to distinguish among many targets in a P300 paradigm, and for that reason these EPs are mostly exploited in building BCI spellers [57]. Despite this advantage, P300 was not considered in the present work, but it could be taken into account in a near future to furtherly investigate hybrid paradigms.

As further relevant measurand signals, SMRs comprise rhythms μ and β, which are oscillations in the brain activity localized in the mu band (7.0 Hz to 13.0 Hz), also known as the Rolandic band, and beta band (13.0 Hz to 30.0 Hz), respectively [49]. The amplitude of the SMRs varies when cerebral activity is related to any motor task although actual movement is not required for the modulation. Similar modulation patterns in the motor rhythms are produced as a result of mental rehearsal of a motor act without any overt motor output. SMRs have been used to control BCIs, because people can learn to generate these modulations voluntarily [49]. The study in [52] proved that, through the only imagination of the movement (motor imagery), it is possible to start variations in SMRs like the ones associated with real movements [49]. Sensory stimulation, motor behavior, and mental imagery can change the functional connectivity within the cortex and results in an amplitude suppression, the *Event-Related Desynchronization* (*ERD*), or in an amplitude enhancement, the *Event-Related Synchronization* (*ERS*), of mu and central beta rhythms. The dynamics of brain oscillations associated with sensory and cognitive processing and motor behavior can form complex spatio-temporal patterns. Thus, a synchronization of higher frequency components embedded in a desynchronization of lower frequency components can be found on a specific electrode location at the same moment of time [58]. Voluntary movement induces ERD of μ and β sensorimotor rhythms. Desynchronization begins in the contralateral hemisphere 2 s before the motor act, then it becomes symmetric during the execution of the movement. The same ERD can be revealed during motor imagery. ERD is often followed by ERS in the ipsilateral hemisphere, with similar frequency components. One important feature of these beta oscillations is their strict somatotopic organization in MEG and EEG [58]. For this reason, patterns associated with the imagination of the movement of a hand can be distinguished with respect to the ones associated with the movement of the other hand. To properly record ERD and ERS, the EEG electrodes must be located close to the primary sensorimotor areas. Usually, the most interesting electrodes for these studies are *C3* and *C4*. However, two bipolar derivations are insufficient to describe the overall brain activity, therefore it seems reasonable to assume that more EEG signals recorded over sensorimotor areas, which are sensitive to differences between left and right imagery, would improve the classification accuracy of the BCI [59]. Furthermore, although electrodes close to primary sensorimotor areas contain the most relevant information for discrimination, surrounding electrodes over premotor and supplementary motor areas also contribute some information to discriminate between brain states related to the motor imagery task [59]. Obviously, the more signals are recorded, the less BCI systems are portable because of the great number of electrodes and because of proportionally longer preparation times. SMRs are utilizable for the design of endogenous BCIs, which are more useful than exogenous BCIs. Nevertheless, self-control of these rhythms is not easy, and most people have difficulties with motor imagery. People tend to imagine visual images of related real movements, which is not sufficiently useful for a BCI system, because the patterns of these sensorimotor rhythms differ from actual motor imagery. SMRs have been investigated extensively in BCI research. Well-known BCI systems such as Wadsworth, Berlin, or Graz BCIs employ sensorimotor rhythms as control signals [49].

1.6 PARADIGMS: REACTIVE, PASSIVE, AND ACTIVE

Zander and al. proposed a distinction between *reactive, active*, and *passive* BCIs [26]. In a *reactive BCI*, the brainwaves are produced in response to external stimuli, and this clearly coincides with the "exogenous BCI" case. Stimuli can be tactile, visual, auditory. It is also worth saying that the subject can consciously or unconsciously be exposed to the external stimuli, and this implies a further distinction between a reactive BCI for control applications versus a monitoring application. The main advantage of this paradigm is that no training period is required for the subjects. The need for the presence of external stimuli, even to provide control commands, is the main disadvantage of reactive BCI. Example of reactive BCI concerns communication applications (e.g., the P300 Speller) or providing control commands.

Secondly, in a *passive BCI* the user does not directly and consciously control his/her electrical brainwaves. Such paradigms are generally used for monitoring the user's mental state. In particular, the main application consists in monitoring the cognitive or emotional state of the subject while performing tasks. In recent years, however, clinical applications for brain activity-based control of games (neurofeedback treatment) are also being developed.

Finally, in an *active BCI* the subject voluntarily produces a modulation of the brain waves for controlling an application, independently of external events. This paradigm is linked to the subject's active performance of cognitive tasks, such as motor imagination. In this type of task, the subject is asked to imagine movements. Participants are required to undergo a period of training in order to learn to imagine movements correctly (i.e., so that the system can classify them correctly). These movements are translated into commands for external devices. The training period can last up to several weeks. The active and passive BCI paradigms can be associated with the "endogenous BCI" category, but of course some aspects of them are better focused.

2

Design of Daily-Life Brain-Computer Interfaces

In this chapter, the acquisition part of a BCI system is treated by considering four common trends of the market. A particular focus is made on electrode types (Section 2.2), portability, low cost and user-friendliness (Sections 2.3 and 2.4) of the architecture. Finally, privacy issues are briefly addressed (Section 2.5).

2.1 ACQUISITION SYSTEM

In designing the signal acquisition for neural activity, the hardware is mainly concerned. Key aspects for the proposed architectures are indeed high wearability and portability of the head-mounted device, and this is especially feasible by choosing EEG as acquisition technique (see Chapter 1). As a further aspect, also EEG non-invasiveness has already been indicated as essential for everyday-life usability. In addition, user-friendliness is enhanced by adopting a limited number of dry electrodes to place on the scalp. This is in contrast with more classical approaches, where many EEG electrodes are exploited and conductive gels are used to provide a good skin-electrode contact. As an example of an actual implementation, a low-cost acquisition system relying on components off the shelf is firstly described. In detail, the Olimex EEG-SMT is an acquisition board with up to two differential channels [60] providing conditioning for the signals recorded by electrodes. For each channel, two active electrodes are connected to the positive and to the negative input terminal, respectively. These electrodes are active since a buffering circuit is integrated onto the electrode, so to enhance impedance matching between electrode and skin (scalp). Meanwhile, a passive electrode without buffer must be connected to the reference input, which is shared by both channels and serves for common mode noise rejection. Figure 2.1 shows this acquisition channel in single-channel configuration.

The signals acquired through the electrodes are amplified and filtered before being digitized. According to the circuit schematic by Olimex Ltd, the nominal bandwidth is 0.16 Hz to 59 Hz, while the nominal gain equals

DOI: 10.1201/9781003263876-2

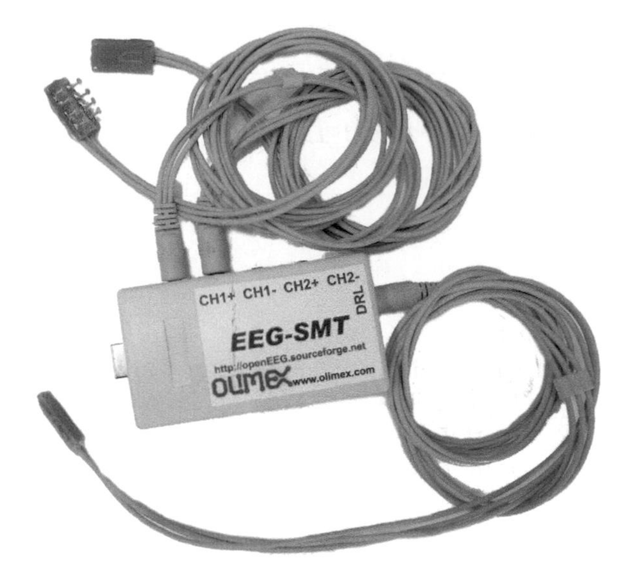

FIGURE 2.1 The Olimex EEG-SMT acquisition board with two active dry electrodes connected to channel 1 (CH^+, CH^-) for a single-channel differential acquisition and a passive dry electrode connected to "drive right leg" (DRL) reference terminal [61].

6427.2 V/V assuming that the gain of the second stage is set at 40 V/V through the trimmer (see [60]). After signal conditioning, the ATMega16 microcontroller provides the analog-to-digital conversion. The resulting 10-bit codes are finally sent through *Universal Asynchronous Receiver-Transmitter* (*UART*) to an external device, for example a personal computer. Note that when only one differential channel is considered (e.g., CH1), the other channel should be short-circuited. Also note that silver pins were added to one of the active electrodes (noticeable in Fig. 2.1). This helps to overcome the hair and reach a suitable connection to the scalp during EEG acquisition. Conversely, pins are not necessary for the other electrodes placed on hairless parts of the scalp/body.

The acquisition system has been here introduced in a single-channel configuration for the sake of simplicity, and this configuration was indeed used in recording brain activity related to SSVEPs. Nonetheless, the double-channel configuration was also employed when trying to acquire motor imagery-related brain activity. Despite that, in motor imagery applications two channels are generally too few. Therefore, as a meaningful example, the *Helmate* by ab-medica [62], a prototypical EEG cap that going to be commercialized soon, is considered here. This wireless device (Fig. 2.2) exploits ten dry electrodes to provide up to eight single-ended channels. The electrodes are dry, made of conductive rubber, and they have Ag/AgCl coating. Three different electrode shapes can be installed for an optimal contact with the scalp, and a

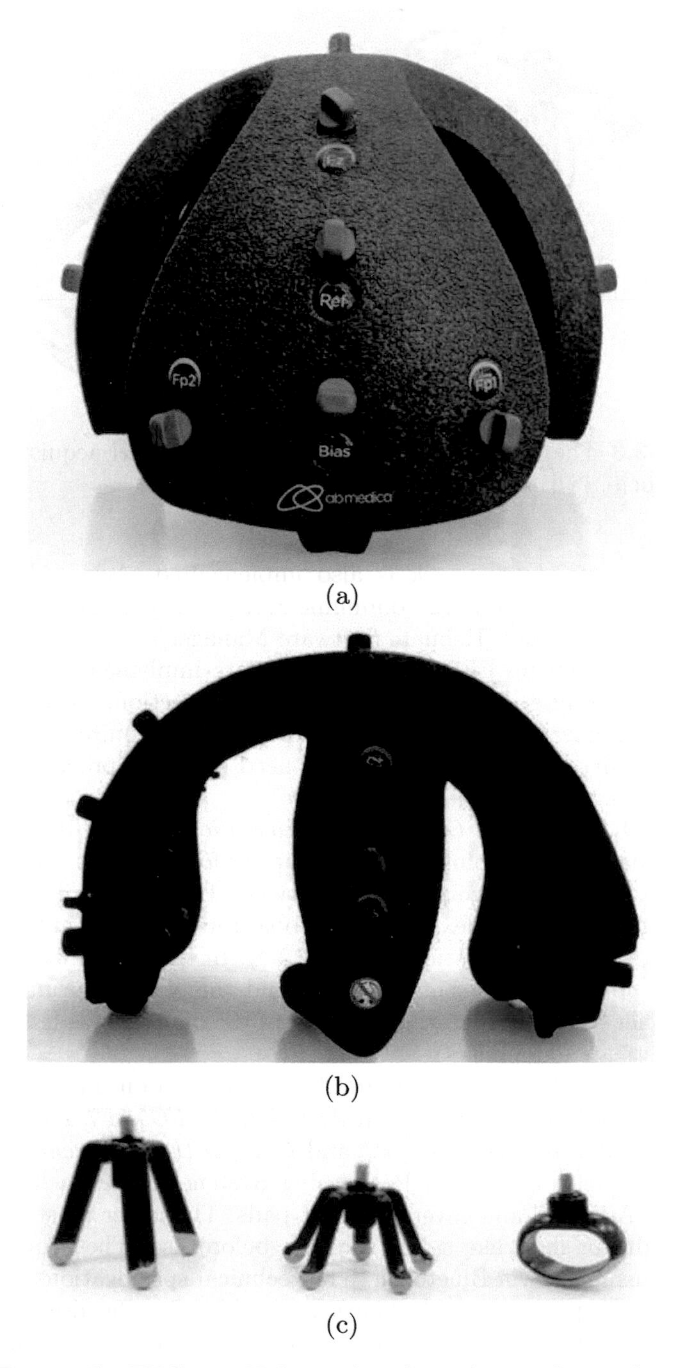

FIGURE 2.2 The EEG cap Helmate by ab medica with ten dry electrodes in different shapes and Bluetooth connection to ensure high wearability and portability: (a) front view, (b) side view, and (c) electrodes [62].

FIGURE 2.3 The wearable, high resolution, 14-channel acquisition system Emotiv Epoc+. [64]

skin-electrode impedance check is also implemented. Acquired signals are transferred through Bluetooth communication to a personal computer with the proprietary software "Helmate Software Manager".
Hence, for the Helmate EEG cap, the hardware implementation details are not available because of intellectual property protection. In a similar vein, also the software is non-free and it is only possible to acquire signals for post-processing analyses, in contrast with the need of online processing for some applications (e.g., neurofeedback).

On the other hand, the *Olimex EEG acquisition system* [61] is open hardware. This makes such a solution very attractive for designers, since they have full control of hardware, firmware, and software. Further details about the internal structure of the Olimex acquisition board will be given in the context of its metrological characterization (Section 2.4). In that framework, there will also be the opportunity to better understand the working principles of this EEG amplifier as a representative example of the main principles shared with more sophisticated amplifiers.

Emotiv Epoc+ (Fig. 2.3) is a wearable, high-resolution, 14-channel acquisition system. Electrodes are placed on *AF3, F7, F3, FC5, T7, P7, O1, O2, P8, T8, FC6, F4, F8, and AF4*, (*P3/P4* and *CMS/DRL* as *reference electrodes*), according to the International Positioning System 10-20. Each electrode is coated with Ag/AgCl and covered by felt pads. The latter must be drenched with an hydrator fluid like saline solution before use. The acquired signals are then transmitted via Bluetooth. The technical specifications of the *Emotiv Epoc+* are reported in [63]. The Emotiv EPOC + headset outputs 128 samples per second and has a 14 bits resolution (0.51 μV step for 8400 μV dynamic range).

Lastly, the *FlexEEG system from Neuroconcise* is here introduced [65]. This wireless device uses Bluetooth communication to guarantee portability other than wearability. It consists of a flexible electronic board for signal conditioning and transmission (Fig. 2.4a), and two possible electrodes configurations can be exploited for visually evoked potentials and motor imagery, respectively. The sensorimotor electrode array, shown in Fig. 2.4b, has been considered hereafter for experiments with motor imagery. It comprises seven star-shaped electrodes (Fig. 2.4c) for a configuration with three bipolar channels, i.e., *CP3-FC3* and *CPz-FCz* and *CP4-FC4*, as well as a *reference electrode* at the *AFz* standard location. These placing allows to map the sensorimotor area of the scalp. The usage of conductive gel is suggested to ensure a proper skin-electrode contact during EEG measurement. In standard settings, electrical brain activity is sampled at 512 Sa/s with 16-bit resolution. Data are received in Simulink and online processing is also possible. In terms of user comfort, the only disadvantage is the need of conductive gel, though this solution was successfully employed in neurofeedback experiments with motor imagery, as discussed in Chapter 10.

2.2 ELECTRODES

The term "electrode" is often misused because it is referred both to the conductor and to the semi-element, strictly speaking in electrochemistry. In the former case, the electrode is a conductor of the first kind, such as metal or graphite. It is used to establish an electrical contact with a non-metallic part of a circuit, for instance a semiconductor or an electrolyte. In the latter case, the electrode is an half-cell including the abovementioned conductor, and namely the interface between the electronic and the ionic conductor.

In this book, the first definition as a conductor for EEG signal measurement, is considered. In the following, Section 2.2.1 reviews the basic mechanisms involved in the transduction process, Section 2.2.2 introduces the equivalent circuits based on electrical properties of the electrodes, and Section 2.2.3 discusses the advantages and disadvantages of using wet or dry electrodes in the measurement of bio-potentials.

2.2.1 Electrode-Electrolyte Interface and Polarization

Phenomena occurring at the electrode-electrolyte interface (Fig. 2.5) concern the transition from ionic to electronic current. They are analyzed hereafter to understand how they occur in wet (with conductive gel) and dry (no conductive gel) electrodes [66]. Biological tissues, sweat and moisture work as electrolyte solutions, together with the eventual conductive gel.

The oxidation-reduction reactions occurring at the interface can be described as:

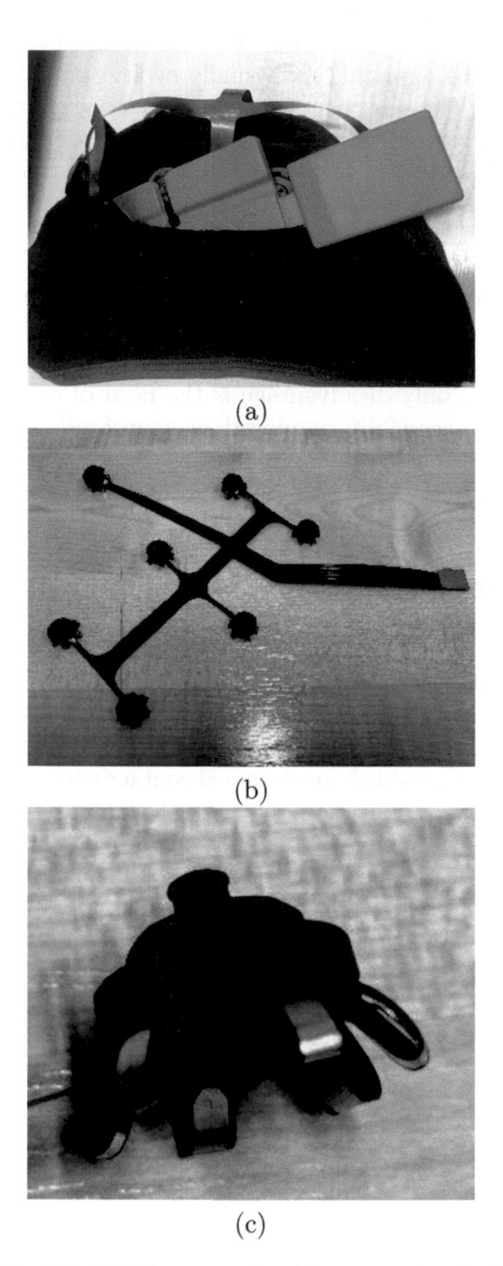

FIGURE 2.4 The FlexEEG system by Neuroconcise Ltd with seven electrodes needing conductive gel. Bluetooth connectivity ensures portability other than wearability: (a) flexible board and electrode array, (b) sensorimotor electrode array, and (c) star-shaped electrodes [65].

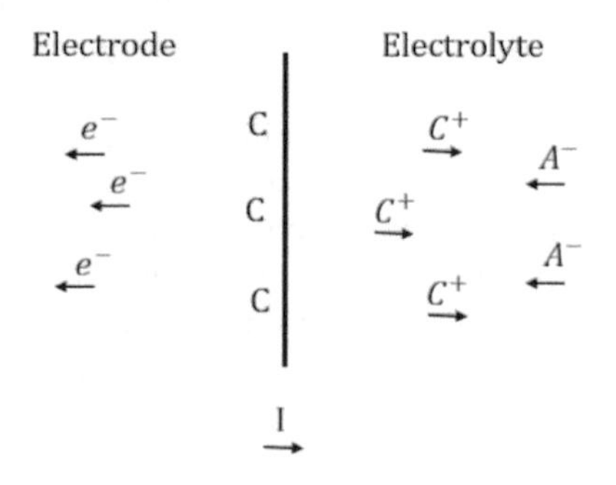

FIGURE 2.5 Electrode-electrolyte interface; the electrode consists of metal C atoms, while the electrolyte solution contains cations of the same metal C+ and A anions. I is the current flow.

$$C \rightleftharpoons C^{n+} + ne^{-12} \tag{2.1}$$

$$A^{m-} \rightleftharpoons A + me^{-} \tag{2.2}$$

In there, the cation is released into the electrolyte solution, while the electron remains in the electrode as described by Eq. (2.1). Eq. (2.2) describes the phenomenon whereby an anion reaching the electrode-electrolyte interface can oxidize to a neutral atom, releasing one or more free electrons into the electrode. In this scenario, there is a current flowing through the interface consisting of (i) electrons, in the electrode, and (ii) anions, in the solution, moving in the opposite direction to the current. Cations in the electrolyte solution move in the direction of the current.

As soon as the metal comes in contact with the solution, the reaction described by Eq. (2.1) begins. The oxidation or reduction reaction will predominate depending on the concentration of cations in solution and the equilibrium conditions. Near the interface, the local concentration of cations, and consequently of anions, changes leading to a non-neutrality of charge: the electrolyte near the metal assumes a different potential value from the rest of the solution. At equilibrium (i.e., when the rate of oxidation reactions is equal to that of reduction, so no current flows at the interface), a separation of charges occurs at the interface leading to the formation of an electrical double layer: charges of one type (positive or negative) accumulate on the surface of the metal, while opposite charges are distributed in excess in the electrolyte solution. This double layer is responsible for a potential difference between electrode and electrolyte, called half-cell potential, to balance the initial motion of the charges. The half-cell potential refers to no current flowing at the interface situation.

When current flows at the interface, the value of the potential changes due to the variation in the distribution of ions in the electrolyte solution at the interface. Theoretically, there are two types of electrodes: polarizable and non-polarizable [67]. In the former, no real charge crosses the interface when a current is applied. Polarizable electrode behaves as a pure capacitor and the half-cell potential changes are significant. In non-polarizable electrodes, the current flows freely across the interface without changing the half-cell potential: there is purely resistive behavior at the interface. In polarizable electrodes, the relative movement between electrode and electrolyte interrupts the charge distribution at the interface. As a result, the half-cell potential varies until equilibrium is restored. In the case of a pair of electrodes in contact with the electrolyte, if one moves relatively to the electrolyte, while the other remains stationary, a potential difference is created between the two electrodes. This potential difference is known as motion artifact and it can be a serious cause of noise and errors in biopotential measurements.

In view of the electrode-electrolyte interface and polarization considerations, the electrode equivalent circuit, suitable for biomedical application, is described in Fig. 2.6. E_{hc} is the half-cell potential, C_d is the capacitance of

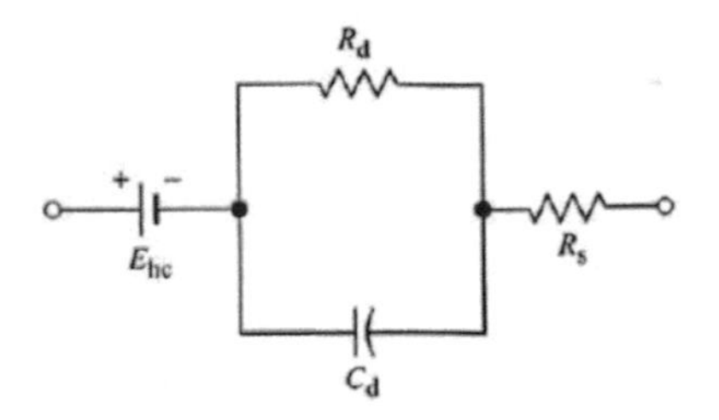

FIGURE 2.6 Equivalent electrode circuit.

the capacitor relative to the electrical double layer (in perfectly polarized electrodes, only this term is present), R_d is the charge transition resistance at the interface, and R_s is the resistance associated with the electrolyte.

2.2.2 Electrode-Skin System

The type of the electrode and the skin interface must be considered in order to fully understand the behavior of the electrode. Indeed, the type of electrode influences the impedance of the electrode and the skin interface. The skin consists of three layers: epidermis, dermis and hypodermis. The epidermis is the outermost layer, and it is the most important for characterizing the interface. The outermost part, i.e., the stratum corneum, is composed of enucleated, keratin-rich cells. The stratum corneum is non-conductive and, therefore, is an obstacle to the passage of electrical current. In addition, the presence of numerous skin adnexa crossing the epidermis (i.e., sebaceous glands, sweat glands, and hair follicles) must be considered for the characterization of the electrical properties of the skin. If the coupling with the skin is achieved by means of electrolyte gel, usually rich in Cl^-, the equivalent circuit of the skin-electrode system is illustrated in Fig. 2.7.

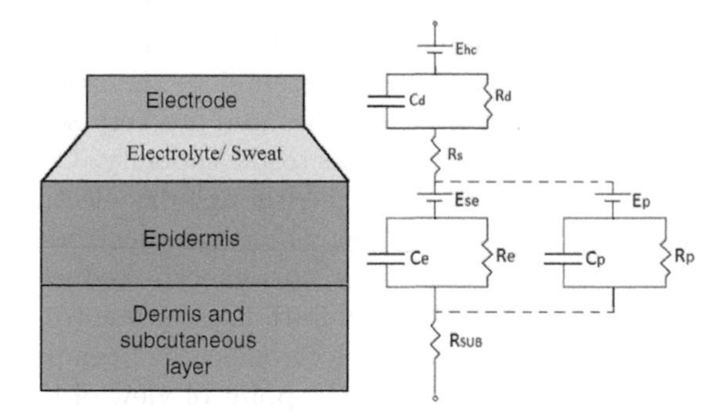

FIGURE 2.7 Equivalent circuit of the wet electrode-skin system [68]. E_{hc} is the half-cell potential, C_d is the capacitance of the capacitor relative to the electrical double layer, R_d is the charge transition resistance at the interface, and R_s is the resistance associated with the electrolyte.

The resistor R_s takes into account the interface effects of the gel between the electrode and the skin. The stratum corneum behaves as a semi-permeable membrane to ions [67]. If the concentrations of ions on either side of the membrane are different, a potential difference E_{se} will be present. In general, the epidermis is modeled by a capacitor C_e and a resistor R_e in parallel.

E_p represents the potential difference between the interior of the sweat ducts and dermis. In fact, the fluid secreted by the sweat glands contains different concentrations of Na^+, K^+ and Cl^- than the extracellular fluid. The parallel $C_p//R_p$ represents the impedance offered by the sweat ducts to the current flow. The resistor R_{sub} represents the resistance offered by the dermis and subcutaneous layer to the passage of current.

2.2.3 Wet and Dry Electrodes

Wet electrodes consist of the electrode and an electrolyte gel to promote good coupling with the skin. The gel reduces the skin-electrode impedance by hydrating the stratum corneum. The characteristic capacity of the epidermis C_e increases and the characteristic resistance R_e decreases. Furthermore, the gel promotes good skin-electrode contact keeping the interface impedance low even during movements. Therefore, there are several advantages of using such electrodes: (i) a high-quality signal and (ii) limited noise due to motion artifacts. However, there are several disadvantages. In particular, before their use, the skin has to be prepared: abrasive gels are used to clean the skin. This operation requires time and an experienced human operator. Moreover, during long-term EEG signal monitoring, there is a progressive degradation of signal quality because the gel starts to dry out. Finally, removal of the electrodes is difficult and time consuming, and skin irritation is often observed. Therefore, dry electrodes were introduced.

Dry electrodes are divided into two macro categories: non-sink contact and direct-skin contact. The non-contact electrodes (e.g, used in touch screen) are capacitively coupled (via displacement currents) to the skin. They provide a very-low amplitude signal at low frequencies and are very sensitive to motion artifacts as motion changes the skin-electrode capacity. Contact electrodes can involve direct or capacitive coupling. Although dry electrodes are simpler to use, they have an impedance higher than wet electrodes. However, for extended use of the wet EEG electrodes, this effect is reduced over time due to sweating acting like an electrolyte, and are more prone to motion artifacts. To overcome this, different types of dry electrodes have been presented over the years that aim to reduce the gap, from the point of view of the signal quality, compared to wet electrodes. Given the lack of electrolytic gel, the equivalent circuit of the electrode-skin system, for dry electrodes, is modified as shown in Fig. 2.8: the interface effects between electrode and skin are modeled as a resistor–capacitor circuit ($C_i//R_i$).

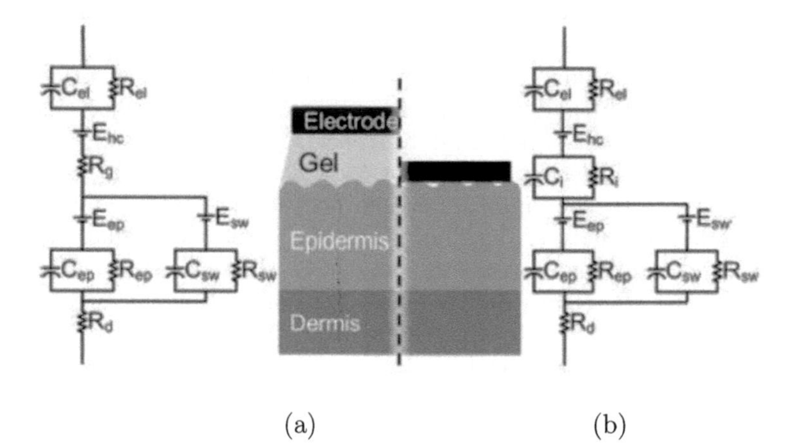

FIGURE 2.8 Circuit model of the electrode-skin system for wet (a) and dry (b) electrodes [69]. E_{hc} is the half-cell potential, C_{el} is the capacitance of the capacitor relative to the electrical double layer, R_{el} is the charge transition resistance at the interface, and R_g is the resistance associated with the presence of the gel. The parallels $C_{ep}//R_{ep}$ and $C_{sw}//R_{sw}$ represent the impedance offered by the epidermis and by the sweat ducts, respectively, to the current flow. R_d represent the resistance offered by the dermis and subcutaneous layer. E_{ep} is the potential difference due to the different concentration of ions on either side of the membrane. E_{sw} represents the potential difference between the interior of the sweat ducts and dermis.

2.3 CHANNEL MINIMIZATION STRATEGIES

The desired characteristics for a daily BCI are portability, non-invasiveness and cost effectiveness. Several studies employed channel minimization strategies, while guaranteeing acceptable performance, in order to increase the portability of BCI solutions in many application fields. In the literature, most studies concerning Motor Imagery-BCI applications proposed channel selection strategies. Some proposals are described below [70]. A research of 2015 [71] took into account 14, 22, and 29 channels. Accuracy results depended on the considered dataset and tasks. Consequently, several datasets must be considered to validate an algorithm and a comparison with the literature must be carried out. In particular, several studies considered the dataset 2a of BCI competition IV [72], [73, 74, 75, 76, 77, 78], proposing different channel minimization strategies.

A *Common Spatial Pattern* (*CSP*) was proposed by Arvaneh et al. [79] to discriminate between left-hand versus right-hand imagery. Accuracies of 81 % and 79 % were achieved by using 13 and 8 channels, respectively. The selected channels varied, however, depending on the subject. A a-priori channel selection was implemented by other studies. In particular, the electrodes *C3, Cz,* and *C4* of the 10-20 system were often considered, being associated to the motor area of the scalp [80]. However, the performance was not high [79, 81]. Also single-channels applications were proposed [82, 83].

Also in emotion recognition, many studies focused on the identification of the minimum number of EEG channels, while keeping high the signal classification performance. The channel reduction strategies mainly employed in the literature can be divided into three categories of studies using: (i) statistical algorithms to identify the best subset of channels; (ii) manual choose based on the neurophysiological theories of the origin of emotions to identify the best subset of channels; and (iii) channel selection resulting from the use of commercial EEG devices with a low number of channels.

As far as statistical algorithm category is concerned, *mutual information-based* and *correlation-based* methods are the mainly implemented for the selection of EEG channels with higher information content. A correlation-based method was applied by MK Ahirwal et al. [84]. They computed the functional inter connections between electrodes pairs and concluded that *CP1, O1, Pz, Po4* are the best electrodes. Another correlation-based method is the Reverse Correlation Algorithm, in which the correlation between one electrode and all the others is computed, for each electrode and summed. Channels with lower value are selected. Finally, mutual information-based methods consist in selecting the low number of channels by evaluating mutual dependence (amount of information) between them. The method was tested by H Xu et al. [85] reducing from 32 to 22 channels with a small loss in classification accuracy.

Neurophysiological-based studies focused mainly on electrodes placed in the frontal area. Other studies considered few electrodes symmetrically from frontal, parietal, temporal, and occipital area of each hemisphere. Finally, in several studies, commercial EEG device are used. In these cases, the choice of channels used depends on the device selected.

2.4 CHARACTERIZATION OF LOW-COST ELECTROENCEPHALOGRAPHS

An electroencephalograph is a specialized voltage digitizer monitoring the electrical activity of the brain. Peak-to-peak amplitudes in normal EEG typically range from 0.5 µV to 100 µV and the frequency band of interest is 0.5 Hz to 100 Hz [28], though the exact band depends on the considered application. Especially for off-the-shelf components, there is the need to calibrate the gain of an EEG acquisition system at different nominal frequencies. Nowadays, in clinical applications, some guidelines exist for EEG calibration prior to the measurement of brain activity [86]. However, there is a lack of standardization in assessing the metrological performances of EEG instruments. These are often validated by acquiring the brain activity of an user with eyes open, or closed, or even by evoking brain potentials [87, 88]. Nonetheless, merely validating by means of evoked potentials could be not appropriate since the human response to stimulation is not yet fully understood (e.g., see [89] for SSVEP). In this context, the *IEEE Standards Association* has indicated a roadmap for neurotechnologies by considering standardized calibration procedures in addition to validation based on user tasks [90]. The work reported hereafter aimed to give a small contribution to the topic. An EEG calibration procedure has been proposed to assess errors and uncertainties related to the instrument, and also to give basic traceability to the international system of units. In doing that, one can exclude or take under control errors in the EEG measurements so as to achieve valuable results with the acquired data.

Owing to its extreme simplicity, the Olimex EEG-SMT plus dry electrodes [60] (Fig. 2.1) is considered in this section. The use of the instrument to acquire electroencephalographic signals from a single differential channel, in the SSVEP-BCI case, is highlighted. Also its usage with two bipolar channels for the motor imagery-BCI is illustrated. Particularly for the SSVEP detection, the band of interest spanned from 1.0 Hz to 60.0 Hz. In processing EEG data, the focus was on amplitude spectra, while the phase response was not considered. Therefore, only asynchronous measurements are considered for the calibration, while synchronous measurements are eventually addressed in future works.

2.4.1 Experiments

The internal structure of the electroencephalograph under test is similar to a typical EEG amplifier, but it was simplified by the manufacturer to be less cumbersome and low cost. It consists of an instrumentation amplifier with two differential channels followed by further filtering and amplifying stages. Active electrodes are connected to the positive and negative terminals of each channel. Meanwhile, a passive electrode must be connected to the *Drive Right Leg* (*DRL*) input to act as a reference potential for common-mode rejection. The distinction between active and passive electrodes is that the former have an operational amplifier-based buffer to improve the electrode-skin contact impedance, while the latter are simply conductors. Passive electrodes could also be employed as input to the differential channels, but this is not recommended for the sake of signal quality. After amplification and filtering, the signals are digitized with an ATMega16 microcontroller: this provides the timing for the ADC, and the resulting 10-bit codes are sent through *Universal Serial Bus* (*USB*) to an external device, such as a personal computer. The sampling rate was set at 256 Sa/s. In the present case, only one differential channel (CH1) was considered, while the other (CH2) was internally short circuited to avoid cross-talk noise. Figure 2.9 shows a part of the EEG acquisition board devoted to signal conditioning before the ADC. Note that a trimmer can be used to adjust the variable gain of the second amplifying stage. In the calibration, the trimmer is set to 1.66 kΩ to achieve a gain $G_2 = 38.6\,\text{V/V}$. According to nominal specifications, the bandwidth should go from 0.16 Hz to 59 Hz, while the nominal gain should equal 6202.25 V/V (given the value set for G_2). Nonetheless, calculations conducted thanks to the circuit schematic already highlighted a gain that was 20 % higher (about 7510 V/V).

The gain of the EEG-SMT was calibrated trough a calibration setup adapted from a lock-in amplifier calibration application [92]. This setup was designed and realized at the Italian national institute for research in metrology ("Instituto Nazionale di Ricerca Metrologica"). This is represented in Fig. 2.10. The signal generator SG provides both (i) a large-amplitude, low-distortion sine wave V_S and (ii) an isofrequential reference square wave V_{REF} for triggering a calibrated voltmeter. Two output channels are employed. This generator is based on the PCI-6733 board from National Instruments. The signal V_S is buffered by a unity-gain amplifier (based on INA111) with differential input, thus providing electrical decoupling from SG as well as the drive current for the following stages. The buffered voltage is fed to a calibrated voltmeter, the Hewlett-Packard 3458A (used in synchronous sub-sampling mode), which is externally triggered by V_{REF} in order to provide an accurate measurement of V_S. Meanwhile, V_S is also scaled by means of cascaded *Inductive Voltage Divider* (*IVD*) and *Resistive Voltage Divider* (*RVD*) stages. The measured scaling factor for IVD(k_{IVD}) goes from 0.0 to 1.0 with 50 ppm relative

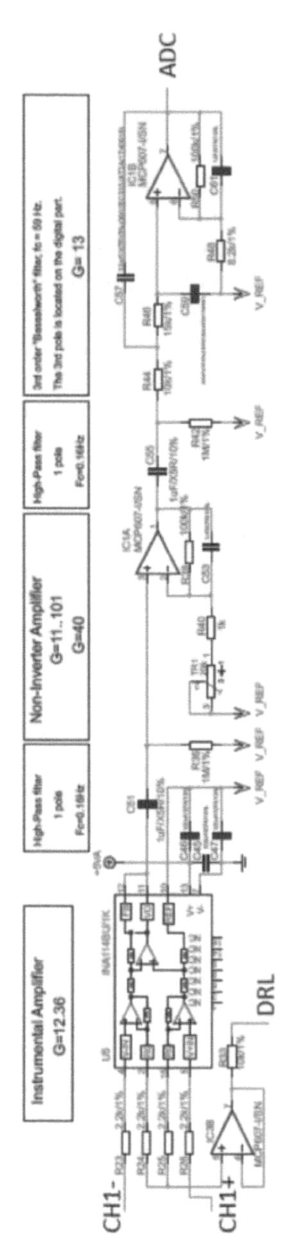

FIGURE 2.9 Signal conditioning circuit being part of the Olimex EEG-SMT acquisition board. The connections to the CH1 pins and the DRL are shown, as well as the input ADC line. Redrawn and adapted from the datasheet of e Olimex EEG-SMT bio-feedback board [91].

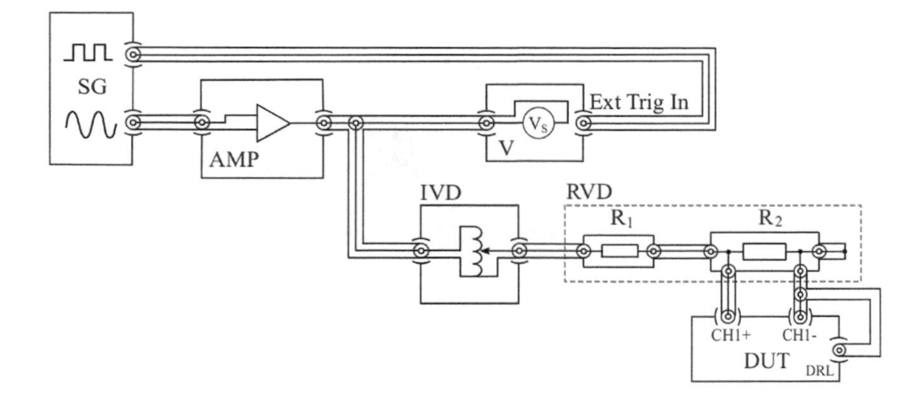

FIGURE 2.10 Coaxial schematic diagram of the EEG calibration setup realized at the "Instituto Nazionale di Ricerca Metrologica".

uncertainty, while for the RVD the measured scaling factor is

$$k_{RVD} = \frac{R_2}{R_1 + R_2} = 9.999317 \cdot 10^{-5} \quad (2.3)$$

with relative uncertainty equal to 13 ppm. The calibration voltage in input to the EEG-SMT, applied between $CH1^+$ and $CH1^-$, results

$$V_{CAL} = k_{IVD} k_{RVD} V_S. \quad (2.4)$$

The instrumentation employed in realizing the calibration setup is reported in Fig. 2.11, together with a detail of the connection to the EEG electrodes. The input voltages are digitized and the output codes are sent through USB to a PC with MATLAB®, which acquires the data in order to process them. The connections to the electrodes are realized with a copper plate and conductive carbon tape. In this way, CH1 is connected to the terminals of the resistance R_2, while the DRL electrode is connected to the generator common, which is also equipotential to the shield of the coaxial cable connected to the generator itself, as shown in Fig. 2.10.

2.4.2 Data Analysis

The acquired codes were scaled according to

$$V_{EEG}(K) = \frac{KQ - SH}{GA}, \quad (2.5)$$

where K is an acquired code, $SH = 2\,\text{V}$ is the *internal level shifting*, $GA = 7509.7\,\text{V/V}$ is the *conditioning circuit nominal gain* calculated from

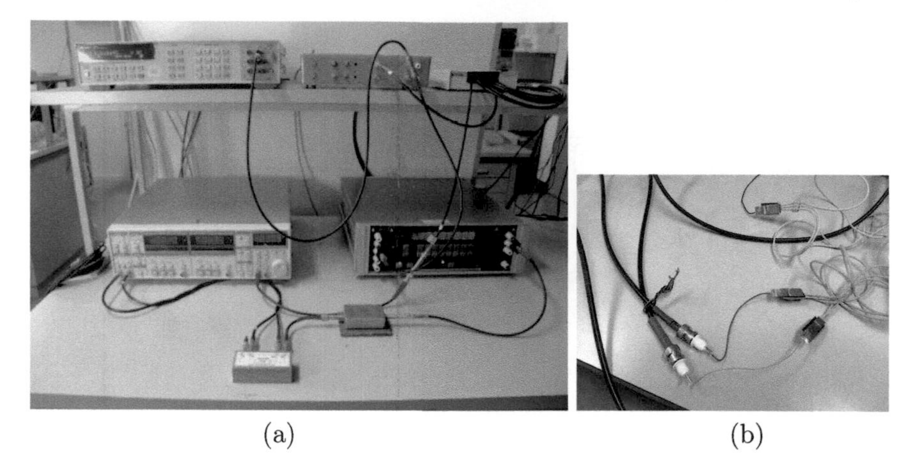

(a) (b)

FIGURE 2.11 Experimental measurement setup for the EEG calibration: (a) instruments connected according to the coaxial scheme of Fig. 2.10, and (b) detail of the connection to the electrodes.

the Olimex Ltd schematic, and the ADC resolution Q is calculated with the *nominal full scale range value* $FS = 4\,\mathrm{V}$ and the *nominal bit resolution* $N = 10$:

$$Q = \frac{FS}{2^N} \tag{2.6}$$

According to the IEEE-1057 standard (see Sec. 4.6 in [93]), the sinefit algorithm is applied to fit the acquired waveforms. In particular, the four-parameter method is exploited: this calculates the values of amplitude (A), offset (O), phase (φ), and frequency (f) of a sinus that give the best fit, in the least squares sense, to the recorded signal. Hence, the sinefit algorithm fits a function of the form

$$V(t_n) = A\cos(2\pi f t_n + \varphi) + O. \tag{2.7}$$

The 4-parameter sinefit is actually an iterative procedure in which an initial guess is needed for the signal frequency. Per each signal to fit, the frequency guess was set equal to the respective nominal frequency set on the generator. A useful function for the fit is available online by MATLAB [94]. It can be implemented according to the above referenced standard IEEE 1057. The root means square (rms) values can be also calculated for both the V_{EEG} and the calibration signal. In particular

$$V_{EEG,rms} = \sqrt{\frac{A_{EEG}^2}{2} + O_{EEG}^2}, \tag{2.8}$$

while the rms of V_{CAL} was calculated by scaling the rms of V_S, namely

$$V_{CAL,rms} = k_{IVD} k_{RVD} \sqrt{\frac{A_S^2}{2} + O_S^2}. \tag{2.9}$$

Indeed, the concept behind the present calibration method is to generate a large and highly accurate signal and then scale it down to the measurement range of interest. Such an approach is reflected into the Eq (2.9) employed for data analysis, and this allows to obtain a great accuracy and a well-defined traceability, in contrast with a case in which one would directly generate a small calibration signal [92].

The uncertainties associated with $V_{EEG,rms}$ and $V_{CAL,rms}$ were also calculated. To this aim, uncertainties had to be estimated for the four parameters of the sinefit. Firstly, the uncertainty associated with the frequency was estimated by considering that the 4-parameter sinefit iteratively employs a 3-parameter sinefit to find A, O, and φ while progressively adjusting f. Starting from the guess value, the frequency is increased or decreased by a fixed amount in each iteration, so as to reduce the error between measured samples and the fitted sine. The iterative procedure stops when the frequency difference between two adjacent iterations is small enough. In the present case, the stopping criterion is set so that the frequency difference was less than 100 ppm. Reasonably, the relative frequency uncertainty is assumed equal precisely to 100 ppm. After that, the uncertainty of the other parameters can be estimated with the variance of the respective estimators as discussed in [95]. In that work, the author analyzes the statistical distribution, bias, and variance of the coefficient estimators for a 3-parameter fit. Under the assumption of Gaussian distribution for the parameters, the variances associated with A and O were estimated to be equal to

$$\sigma_A^2 = 4\frac{\sigma^2}{N} \quad \text{and} \quad \sigma_O^2 = \frac{\sigma^2}{N},$$

respectively, with σ^2 equal to the sample variance associated with fit residuals. Hence, at the end of the fit, the sample-by-sample difference between the measured data and the fit sine of (2.7) can be used to estimate amplitude uncertainties. Given that, the law of propagation of uncertainties is applied [96] to achieve the type A uncertainty

$$u_{V_{EEG,rms}} = \frac{1}{V_{EEG,rms}} \sqrt{\frac{A_{EEG}^2}{2}\sigma_{A_{EEG}}^2 + O_{EEG}^2 \sigma_{O_{EEG}}^2}. \tag{2.10}$$

An analogous expression is obtained for the type A uncertainty of V_S. Then, the type B uncertainty derived from the specifications of the voltmeter is combined to achieve

$$u_{V_{S,rmsC}} = \sqrt{u_{V_{S,rms}}^2 + \frac{t_{V_S}^2}{3}}, \tag{2.11}$$

where $t_{V_S} = 110\,\mu\text{V}$ is the tolerance resulting from the datasheet for the voltage measurement; by assuming a uniform distribution associated with the tolerance value, it must be divided by $\sqrt{3}$ to achieve a standard deviation associated with a Gaussian distribution. Note that this normalization is mandatory before combining the two uncertainty contributions. Finally, the uncertainty of $V_{CAL,rms}$ is obtained by propagating the uncertainties in (2.9), i.e.,

$$u_{V_{CAL,rms}} = V_{CAL,rms}\sqrt{\left(\frac{u_{V_{S,rmsC}}}{V_{S,rms}}\right)^2 + \left(\frac{u_{k_{RVD}}}{k_{RVD}}\right)^2 + \left(\frac{u_{k_{IVD}}}{k_{IVD}}\right)^2}, \quad (2.12)$$

where the relative uncertainties of RVD and IVD have been also taken into account.

2.4.3 Discussion

The characterization of the EEG device was illustrated at frequency up to 100 Hz with the described setup. Both the linearity and the magnitude error were measured. In addition, frequency errors were detected. Linearity was first assessed through measures conducted with $V_{CAL,rms}$ at 20 Hz and seven different amplitudes: 10 µV, 20 µV, 30 µV, 40 µV, 60 µV, 80 µV and 100 µV. Figure 2.12 shows the result obtained by plotting $V_{EEG,rms}$ as a function of $V_{CAL,rms}$. Clearly, linear behavior is visible. This was confirmed by executing

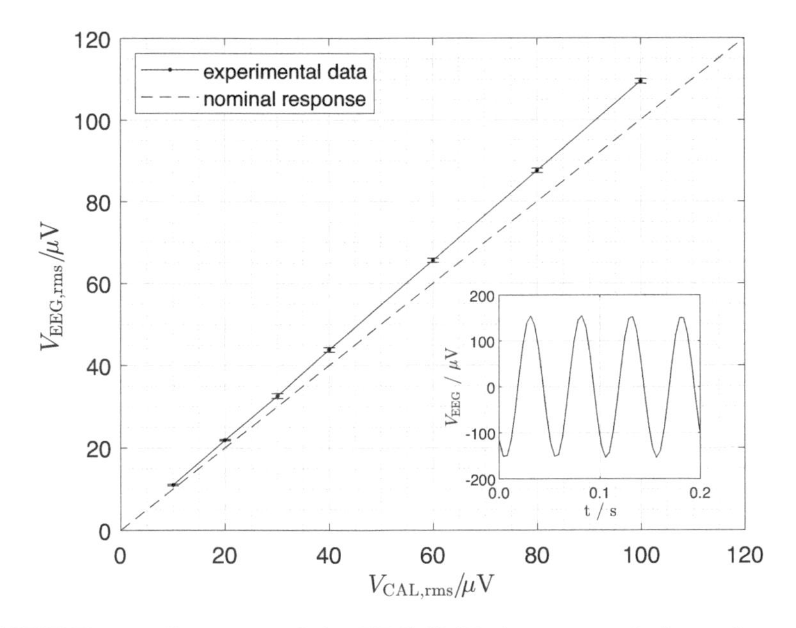

FIGURE 2.12 Linearity of the EEG-SMT electroencephalograph measured at 20 Hz (inset: sine wave acquired with for $V_{CAL,rms} = 100\,\mu V$) [97].

a linear fit and then a Fisher test for the goodness of fitting (p-value $< 10^{-13}$). The ideal response with unitary gain and zero offset is also reported with a dashed line to better highlight linear errors. Notably, the sought magnitude error is actually a gain error, and from these measures it resulted in about 9.5 %. Meanwhile, the offset error resulted in less than −0.08 µV.

The magnitude error as a function of the frequency was calculated according to the expression

$$\epsilon = \frac{V_{EEG,rms} - V_{CAL,rms}}{V_{CAL,rms}}, \tag{2.13}$$

and the associated uncertainty can be calculated by propagating the uncertainty of the two rms voltages according to the law of propagation of uncertainties [96]. By doing that, the obtained curve is reported in Fig. 2.13. The error bars associated with each measuring point represent the propagated uncertainty, expanded with coverage factor $k = 4$. The largest uncertainty on this gain error is 0.3 % and it is associated with 100 Hz.

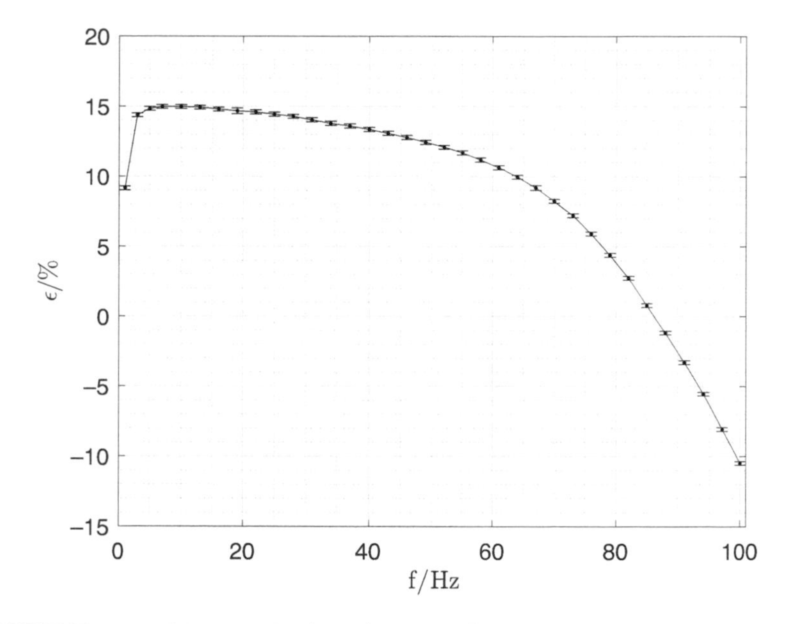

FIGURE 2.13 Magnitude (gain) error of the EEG-SMT electroencephalograph measured at frequencies up to 100 Hz [97].

In explaining the measured gain error, a circuit simulation can be performed thanks to the possibility to replicate the EEG schematic [60], which is open source. The software LTspice® by Analog Devices can be used to set a Monte Carlo analysis. In doing that, nominal values can be assigned to the passive components constituting the conditioning circuitry, and then their declared tolerances were also taken into account. The circuit implemented in LTspice, together with the analysis setting, is represented in Fig. 2.14. The circuit simulation was executed by repeating an AC sweep analysis more than

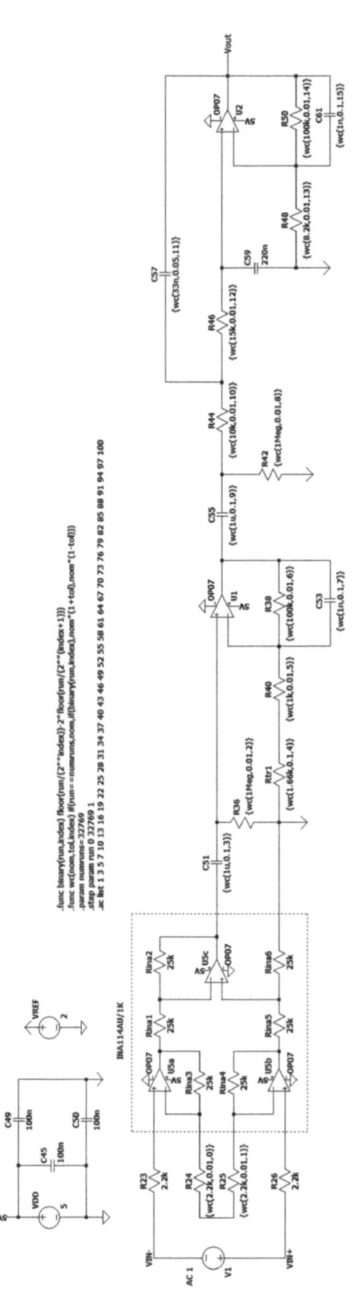

FIGURE 2.14 Analog conditioning circuitry of the Olimex EEG-SMT board replicated in LTspice for executing a Monte Carlo analysis.

32,000 times. In each repetition, a pseudo-random value was selected for each component by considering the set of possible values defined by the respective tolerance. Instead, the AC sweep instead consisted of calculating the amplitude of the output voltage for different frequency values in the range 1 Hz to 100 Hz. Then, the circuit gain was obtained by the ratio of these voltage values and the input voltage, here set at 1 V. Finally, the gain error was finally obtained in relative terms to the nominal gain declared by the manufacturer (7510 V/V). The results are reported in Fig. 2.15 and compared with the measured gain error. The thick gray line corresponds to the simulated gain error when all components assume their nominal value. The gray levels are related to different probabilities of having a certain gain error. In particular, these were arranged in three levels:

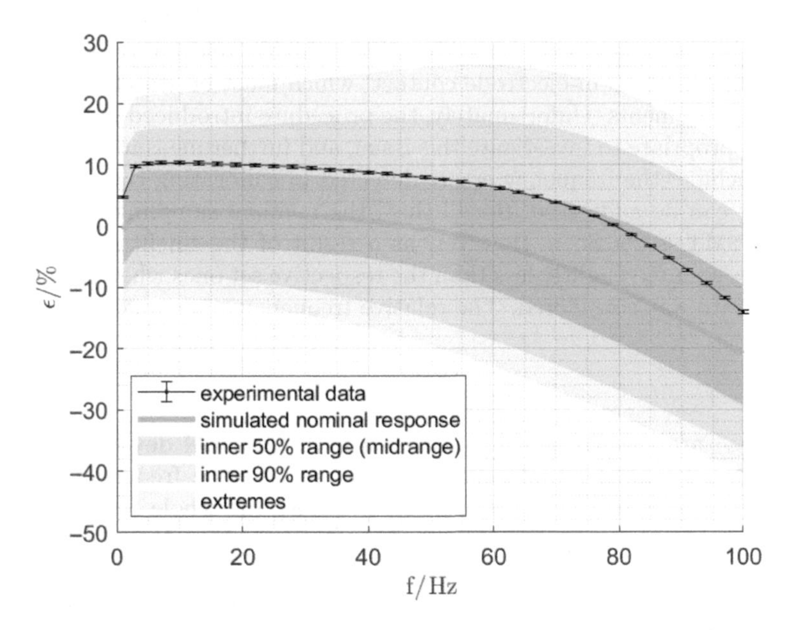

FIGURE 2.15 Simulated gain error obtained with a Monte Carlo analysis (gray line and shaded areas) and compared to the measured gain error (black line with bars). The different gray levels correspond to three different probabilities of occurrence for simulated values [97].

- the inner 50 % range, or mid-range, refers to the interval from the 25th to the 75th percentile and it is represented in dark gray;
- the inner 90 % range, which refers to the interval from the 5th to 95th percentile, is represented in gray;
- the range of all possible values is instead represented in light gray.

Interestingly, the measured gain error lies at the limit of the mid range and it is in principle explained by tolerances associated with circuit components. Nonetheless, the measured curve is almost constantly shifted with respect to the simulated gain error corresponding to nominal values. Therefore, the most probable reason for the 10 % gain error is a drift in the variable gain stage (second stage, i.e., G_2), and calibrating the EEG may be as simple as adjusting the trimmer for setting a different G_2 value.

During experiments with BCIs, the uncertainty associated with the brain signals themselves (intrinsic uncertainty) and the electrodes-scalp contact impedance are usually higher than the measured gain error. However, the gain error can be corrected by calibrating the EEG with the results reported above. The uncertainty associated with this correction is negligible in typical electroencephalographic measures. This implies that even a low-cost device like the Olimex EEG-SMT [61] can properly measure electrical brain activity. Nonetheless, it is worth remarking that a highly relevant aspect in EEG measurement is the skin-electrode contact, which must be stable enough during the measurements. Unfortunately, the procedure introduced in this section was not appropriate to investigate this issue, and further measures are needed.

To conclude, the frequency errors detected in calibrating the EEG device under test can be addressed. Recall that, the nominal sampling frequency of the EEG-SMT is 256 Sa/s. However, as a result of the sinefit algorithm, all frequencies resulted slightly less than the respective set ones when the nominal sampling rate was considered. The relative frequency error was thus assessed by defining

$$\epsilon_f = \frac{f_{EEG} - f_{CAL}}{f_{CAL}}, \tag{2.14}$$

where f_{EEG} are the frequencies measured with the EEG device under test, while f_{CAL} are the nominal generator frequencies. The frequency stability of the generator is in the order of ppm, therefore the whole frequency error should be led back to the EEG-SMT. This error is plotted in Fig. 2.16 as a function of the f_{CAL} values. These errors were reasonably explained with a sampling frequency of the EEG different from the nominal one. In particular, it was derived that the actual sampling frequency equals 257 Sa/s with 1 Sa/s uncertainty. Moreover, instabilities of the internal clock also affect the actual value of the sampling frequency.

2.5 CYBERSECURITY AND PRIVACY ISSUES

Exiting the laboratory environment will indeed be a big change for the BCI field, and its impact on society is still under discussion. Hence, such a contingency poses both technological and ethical issues. From the technological point of view, the scientific and technical communities already understand

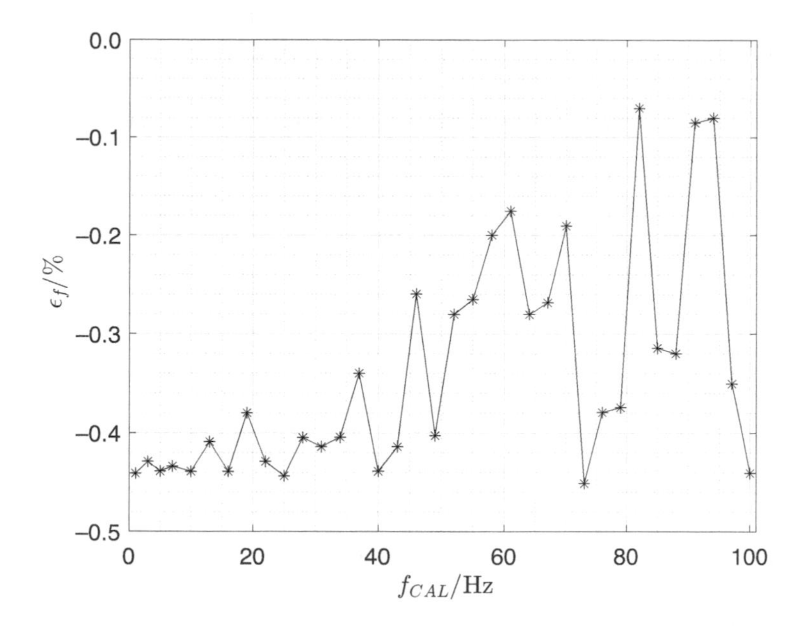

FIGURE 2.16 Relative error between the measured frequencies and the generator ones in percentage.

the need to have standardized measurement procedures, both for acquisition and processing, without which the mere availability of much brain data would not be enough [98]. Unfortunately, the efforts in this direction are still at an early stage. Another interesting initiative has risen in the last few years in trying to also standardize the processing of brain signals and build a common platform to share and compare the results of different research groups [99]. The standardization needs could benefit from that, but today this is still poorly diffused. Beyond that, the possibility to share standardized procedures is still facing an intensive and hard discussion phase. Lastly, the tremendous potential of BCIs clearly poses ethical issues. Although BCI are still very far from "mind reading", privacy issues are already present in acquiring and storing brain data [100], especially when applications like neuromarketing are sought. Hence, ethics will probably follow and guide the development of BCI technologies even more.

In addition, cyber attacks can be carried out to alter the operation of a BCI-based device. From a legislative point of view, new types of crimes and, consequently, new protections to be guaranteed will need to be defined. Ienca and Haselager [101] introduced the definition of *brain hacking* for indicating the act of exploiting the neural device to gain illicit access to brain information and possibly manipulate it in the same way a hacker would operate in cybercrime.

The analysis of brain hackings can be based on the different applications and related electroencephalographic patterns. In (Fig. 2.17) the neurocrimes,

Use case	Brain-hacking	Damage	Countermeasure
Identification (Electroencephalographic signature)	Identity theft	Economical	Encription
Device control based on motor imagery or event related potential	Control system tampering	Physical	Automatic measure for detecting anomalies
		Phsycological	
	Hijacking	Multifactorial	Feedback system to send alarms
Device control based on event related potential	Theft of sensitive data	Social	Neuroattack avoidance training
	Phishing	Economical	
Therapeutic neurofeedback and mental state monitoring	Violation of privacy	Social	
	Control system tampering	Physical	Automatic measure for detecting anomalies
		Phsycological	

FIGURE 2.17 Brain hacking, damages and countermeasures in different BCI use cases.

the damages and the countermeasures are reported, for each use case.

The epoch in which it will be possible to directly access the information contained in the brain continues to approach. Therefore, it is urgent to adopt adequate countermeasures already in the planning phase of future BCIs and to update the regulatory framework. Otherwise, the issues related to forms of brain hacking will become the most relevant for the protection of public safety. Currently, the Authors are engaged in a study activity with the *Research Centre of European Private Law* (*ReCEPL*) [102] aimed at identifying new legal arrangement appropriate to the challenges posed by BCIs.

Part II

Reactive Brain-Computer Interfaces

3

Fundamentals

This chapter focuses on the fundamentals of reactive BCI. In particular, BCI based on SSVEP are analyzed. The main issues concern the detection of the SSVEPs (section 3.1), and the requirements for implementing a daily-life BCI (i.e., wearability, user friendliness, low cost) (Section 3.2).

3.1 DETECTION OF STEADY-STATE VISUALLY EVOKED POTENTIALS

This Section focus on the physiological basis of the signal and its measurement setup of the SSVEP.

3.1.1 Physiological Basis

Evoked potentials are electrical brain signals caused by sensory stimulation. The related electrical patterns reflect brain mechanisms and their measurement can either provide further understanding or an alternative communication way for the user. *VEPs* have been largely exploited in developing EEG-based BCIs [24, 103, 104]. Certainly, investigated paradigms also involve acoustic [105] or tactile stimulation [106], especially when dealing with people with vision impairment. Nonetheless, the visual channel is the most intuitive one and hence the most studied. Different types of VEPs can be distinguished [56]. Notably, *Steady-State Visually Evoked Potentials* (SSVEPs), are elicited by lights flickering so that the effects of consecutive flashes overlap, thus giving rise to a steady state. The minimum flickering frequency for eliciting SSVEPs is about 6 Hz. Meanwhile, an upper limit is not explicitly determined: human response to flickering was studied up to 100 Hz, but most studies do not overcome 60 Hz [104, 107].

The operating principle of SSVEP-based BCIs is rather simple. If the user gazes at a flickering light with frequency f, the evoked potential has fundamental frequency equal to f. Therefore, if multiple lights are flickering at different frequencies, the SSVEP frequency detected from the measured EEG is associated with the specific light the user is gazing at. Actually, different flickering lights could also be distinguished by means of their phase other than

DOI: 10.1201/9781003263876-3

their frequency. Ultimately, SSVEP allows a user to communicate by merely staring at a light or an icon on a display. The advantage is that, at least in principle, no training is required neither for the system operation nor for the BCI user [108]. Despite that, the system can be tuned on a specific subject to enhance detection performances with minimal adjustments. For instance, the processing algorithm can be trained for a specific user with little training, or the electrodes position could be personalized for each user. These aspects will be better discussed next with specific regard to SSVEP-BCI.

3.1.2 Measurement Setup

As mentioned above, the main drawback of reactive BCIs is the need for external stimuli. In the SSVEP case, the previously underlined advantages are thus counterbalanced by the need for the user to stare at a flashing light and keep his/her eyes fixed for a certain time. Nevertheless, many studies showed that SSVEPs are very reliable owing to good reproducibility and superior classification accuracy if compared to other BCI paradigms [109, 110, 111]. The working principle involves shifting the eye gaze, therefore, in line of principle, a SSVEP-based BCI could be replaced by an eye tracker, which has higher performance in recognizing eye position. This appears true when considering that most BCI studies concede eye movements. However, it was shown that, in a SSVEP-BCI, the user can shift attention rather than gaze [112]. So that it is not necessary nor sufficient to move eyes, but the user must focus his/her attention on the flickering light. Given that, SSVEP-BCIs can solve a major issue in eye tracking technology, where the system cannot distinguish an unintentional fixation from intentional ones. Also, SSVEP applications can be addressed to both healthy people and to people with no oculomotor control, such as patients in advanced stages of Amyotrophic Lateral Sclerosis [49].

In this book, Smart Glasses are exploited for visual stimuli generation. Relevant research solutions were resumed in [113] by reporting that the main application field for VEP-based BCIs is robotics, and that video see-through glasses are mostly considered. However, many limitations still prevent the usage of these systems in daily life, especially artifacts [114, 115] and the trade-off between SSVEP detection speed and classification accuracy [103]. These issues are exacerbated when dry electrodes are employed to increase user comfort. Hence, even recent works turn out to be feasibility studies [116, 117]. As a first example of competing SSVEP-BCIs, in 2018, a single-channel BCI was proposed [107] as a speller relying on high-frequency stimuli. Only five subjects participated in the experimental campaign. Each user had to stare at the icon for 10 s and they reached a mean classification accuracy equal to 99.2 %. However, the reproducibility of the results is not assured given the limited number of subjects. As a further example, another speller was proposed in 2019 [118] by exploiting a single acquisition channel, dry electrodes, and a Deep Neural Network for signal processing. The mean classification accuracy among eight subjects was 97.4 %, when each subject starred at the flickering lights for 2 s. Despite the good performance, still a few subjects were involved in the campaign and many trials (about 500) were needed to train the Deep Network.

Finally, a BCI integrated with Augmented Reality (AR) glasses for controlling a quadcopter was proposed by [116]. In this solution, 16 dry electrodes were used, and the mean classification accuracy on five subjects resulted in 85 % while executing a flight task. This accuracy is lower than the previously reported ones, although more electrodes were used. This probably happens because of the flight task that lowers users' attention. Hence, this is a representative example of how leaving controlled laboratory conditions can affect the BCI performance.

3.2 STATEMENT OF THE METROLOGICAL PROBLEM

The use of BCIs as assistive devices for impaired or able-bodied people is becoming increasingly widespread. Many efforts have been made to bring BCI technology out of the laboratory [119], and many attempts are still ongoing to make this possible, but applications in everyday life are nowadays limited due to technological and practical issues. This book would like to contribute to the spreading of such neural interfaces in daily-life activities by providing a more accessible technology. Therefore, the present section first discusses some requirements that a daily-life BCI should have, and then reports some practical considerations devoted to the implementation of such systems.

3.2.1 Requirements

The first requirement for a daily-life neural interface is non-invasiveness. There are two main reasons: the risk of surgical intervention for an invasive device, and its poor acceptance by a large audience of users. Indeed, despite the risks, an invasive BCI could be essential for users with severe disabilities. However, their employment is currently poorly accepted in daily-life usage. Though such considerations may sound trivial, it is worth stressing that the choice of non-invasive neuroimaging techniques limits the brain activity measurement.

Next, wearability and portability are required. Often these two requirements are confused, and the only keyword *wearable* is employed. Actually, also this thesis refers to the treated systems as "wearable Brain-Computer Interfaces" for the sake of brevity. However, it is to remark that a device may be easy to wear but not portable, and still that a portable device/system may be not wearable. For instance, an electronic device acquiring biosignals may be worn on the head, but it could be not portable if it does not have a battery supply independent from the main supply. The wearability and portability requirements comply with the need to leave the user as free to move as possible, and in general the aim is to provide high user-friendliness to avoid fatigue in long-term usage.

These constraints are right away translated into the need for using a lightweight signal acquisition system, as well as into the usage of gel-free electrodes. Notably, these constraints not only affect the hardware part, but they pose some constraints on signal processing too. As an example, motion artifacts diminish the signal-to-noise ratio of the acquired brain signals and proper processing is needed for artifacts removal. Moreover, wearability typically implies the usage of few channels, therefore classification algorithms can exploit less information than usual BCIs.

Another obvious requirement is low cost. Moreover, the BCI system is supposed to exploit as few components as possible to also guarantee easy reproducibility.

The requirements stated above justify the large employment of EEG as brain activity acquisition technique [27], since it is non-invasive and relatively low cost, as well as it can be wearable and portable. For the same reasons, EEG was also considered as a main design choice for the systems discussed hereafter.

As a last aspect considered here, there is the possibility to have minimal training (or ideally no training), namely to avoid long periods in which the system must be tuned before the user can use it. Unfortunately, this aspect is critical because of (i) inter-subject variability, which would require the system to be tuned on each specific user, and even more because of (ii) intra-subject variability, which would require a re-tuning every time the user would like to use the device or even during the usage itself. As discussed later on, such issues are harder to address for active BCIs when compared to reactive BCI paradigms. Indeed, when the paradigm relies on spontaneous brain activity, there is the need of a proper training protocol for the user to learn how to voluntarily modulate his/her brain signals, while the detection of evoked potentials is usually more robust. In brief, this book will suggest that today's reactive BCIs are far more suitable for daily-life applications since they are performant enough even with little user training. Conversely, active BCIs require proper user training protocols and further investigation in transfer learning techniques for addressing algorithm pre-training and signals non-stationarity.

3.2.2 Implementations

The interest in wearable BCI as daily-life assistive devices has been recently increasing. In a recent review on BCIs used in games [15], the progresses in EEG-based devices for video games were reported by showing that the gaming field greatly contributes to orienting BCI issues and concerns. For instance, user motivation is crucial in BCI functionality, therefore games are a good research tool for testing daily-life usage. This review highlighted how BCI are far from applications for healthy people, and that, even though the interest in using commercially available devices is increasing, widespread adoption of such a technology has not yet been triggered. Notably, it was also remarked that motor imagery paradigms are not usually implemented with commercial devices. Among commercial devices, indeed the *Emotiv EPOC* [63] is one of the most exploited in BCI implementations. For instance, a wearable BCI based on SSVEPs was implemented to control a quadcopter [116]. In that work, the attempt was to integrate in a single head-mounted device both the EEG recording system and a virtual-reality visor, by following the idea that the "Smart Glasses" can replace the conventional displays of visual stimuli. This concept will be also adopted in this chapter [31].

Other commercial devices indeed exist as interesting solutions for daily-life applications. Among them, it is worth mentioning EEG caps from *g.tec* [120], which are more devoted to clinical applications, and EEG caps from *Neurosky* [121] and *Bitbrain* [122], whose designs are more prone to user-friendly implementations. Moreover, open hardware solutions are also quite interesting due to the possibility to customize the acquisition system [123]. In these regards, some ongoing studies are trying to assess the quality of signals measured by such commercial devices, and their performance is often compared to EEG systems with wet electrodes, which are de facto standards [124, 29].

3.2.3 Perspectives

Today's society is data hungry. The possibility to have at our disposal a huge amount of data has disclosed new possibilities in many technological fields, such as image processing. Once the *Big Data* are properly stored, which is itself a challenge, their analysis is not trivial, but Deep Learning has demonstrated an extraordinary capability in identifying unimaginable patterns in data, thus giving new insights on many phenomena. Therefore, the interest in adopting novel Deep Learning techniques even in the BCI field is clearly justified. A model identified by means of Deep Learning could give many new insights in the complex behavior of brain activity, thus opening up new possibilities both in terms of more powerful interfaces and more understanding of the human brain. Nonetheless, the application of such techniques to brain signals analysis is still struggling because of the limited amount of structured brain data. Data augmentation and artificial data synthesis are also under study, as well as transfer learning has been recently considered to enable the usage of powerful processing techniques requiring Big Data. In this context, daily-life usage of BCIs would allow to acquire much more real data related to the brain.

3.2.4 Metrological Considerations

In designing a wearable, portable, and relatively low-cost BCI, off-the-shelf components are a suitable choice. In a reactive BCI, such a choice is not exclusively limited to the EEG acquisition, but it is also extended to the stimuli generator. For instance, in the BCI based on SSVEP, Smart Glasses are exploited to generate the flickering lights needed to elicit SSVEP signals. Although many literature studies have been based on commercially available components, the possibility to employ such instruments was not properly justified. Indeed, several studies proved that employing Smart Glasses for generating visual stimuli and commercial or custom EEG for measuring brain activity is feasible, at least from an operational point of view [31, 116, 117]. Nonetheless, off-the-shelf components are much different from laboratory instrumentation, and they could affect the BCI system performance in an uncontrolled manner, if a proper characterization is not carried out.

A metrological analysis of its building blocks is to be conducted to better understand the limiting factors deriving from the usage of off-the-shelf components in a BCI system. The SSVEP-based BCI is here taken into account as case study. This is particularly useful for highlighting the aspects of interest for the characterization. Nevertheless, the electroencephalograph calibration is useful for EEG-based BCIs in general, while the characterization of Smart Glasses applies only to systems requiring external stimuli. Note that characterizing EEG devices by exploiting evoked potentials has been already proposed in literature studies [87, 88]. However, in those approaches, the characterization depends on the subjects' response to the external stimuli because the experiments consist of measuring the brain activity during a user task. In this discussion, instead, each component is characterized separately and independently from the BCI user.

The building blocks of the SSVEP-BCI are represented in Fig. 3.1: Smart Glasses are used for generating flickering icons, the human user transduces this visual stimulation into a SSVEP oscillation, and the EEG device measures the brain activity containing the SSVEP.

The need for a characterization of the components arose from BCI experiments pointing out discrepancies between the expected system response and the measured response. In details, in experiments with SSVEP, the user was stimulated with a flickering light that should have followed a square-wave path (Fig. 3.2(a), but the resulting EEG spectra showed some unexpected harmonic components with unexplained amplitudes (Fig. 3.2(b).

Indeed, the major limitation to the system performance comes from the human user because of the poor signal-to-noise ratio, non-linearity, and non-stationarity of the measurand [125]. In metrological terms, the human vision system could be considered as a transducer with high intrinsic uncertainty. Therefore, in analyzing the whole system, the uncertainty associated with the other two blocks should be negligible. This is particularly true both in laboratory or clinical applications, where it is possible to choose the stimuli generator, and in the signal acquisition system, without caring much about wearability, portability, and/or cost. On the other side, such considerations do not necessarily hold in the consumer-grade context. Therefore, the following sections illustrate analysis carried out for each block separately to quantify

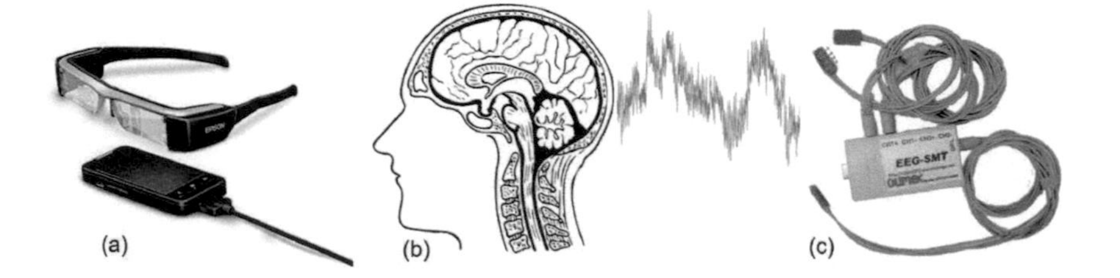

FIGURE 3.1 Wearable BCI system based on SSVEP: (a) the BT200 AR glasses for stimuli generation, (b) the human user, and (c) the Olimex EEG-SMT data acquisition board with dry electrodes [97].

uncertainties and errors introduced by the the EEG acquisition device and the Smart Glasses. This metrological analysis appears essential in justifying the approach adopted in designing wearable interfaces and address future improvements.

3.2.5 Signal Quality

The quality of the recorded EEG signal is a crucial factor for the BCI operation. In particular, the possibility of evaluating the acquisition quality is typically related to the detection of artifacts. Such artifacts can be divided into internal, such as eye blinks, muscular contraction, or heart beat noise, and external, such as electromagnetic interferences, or power line noise.

Indeed, proper EEG electrodes are critical for measuring artifacts-free brain activity. Unstable positioning and deterioration are two important aspects. These issues are worsened in case of dry electrodes, i.e., when conductive gels are not employed in contacting the user's scalp. In this regard, some studies have been conducted to compare different electrodes applications, namely *wet, semi-wet,* and *dry.* Main findings are well resumed in [126]. In there, the magnitude and stability of the electrode-skin impedance was investigated. Reasonably, the contact impedance associated with dry electrodes resulted the highest and most unstable one. This can clearly lead to a distortion of the actual brain signals and, in some cases, no satisfactory EEG can be obtained.

Nonetheless, the results of the review suggest how to lower the electrode-skin contact impedance. First of all, the usage of active electrodes is recommended to improve the measurement quality by pre-amplifying the brain signals as soon as they are picked up. Then, the forearm resulted the worst skin location in comparison with hairy scalp and forehead, so it should be avoided. For instance, one can place the reference electrode on the ear instead of the forearm.

As already mentioned above, silver pins were added to the active electrode of the Olimex EEG-SMT [61] when contacting the skin in hairy scalp areas. A dense set of pins is also suggested because increasing the contact area is

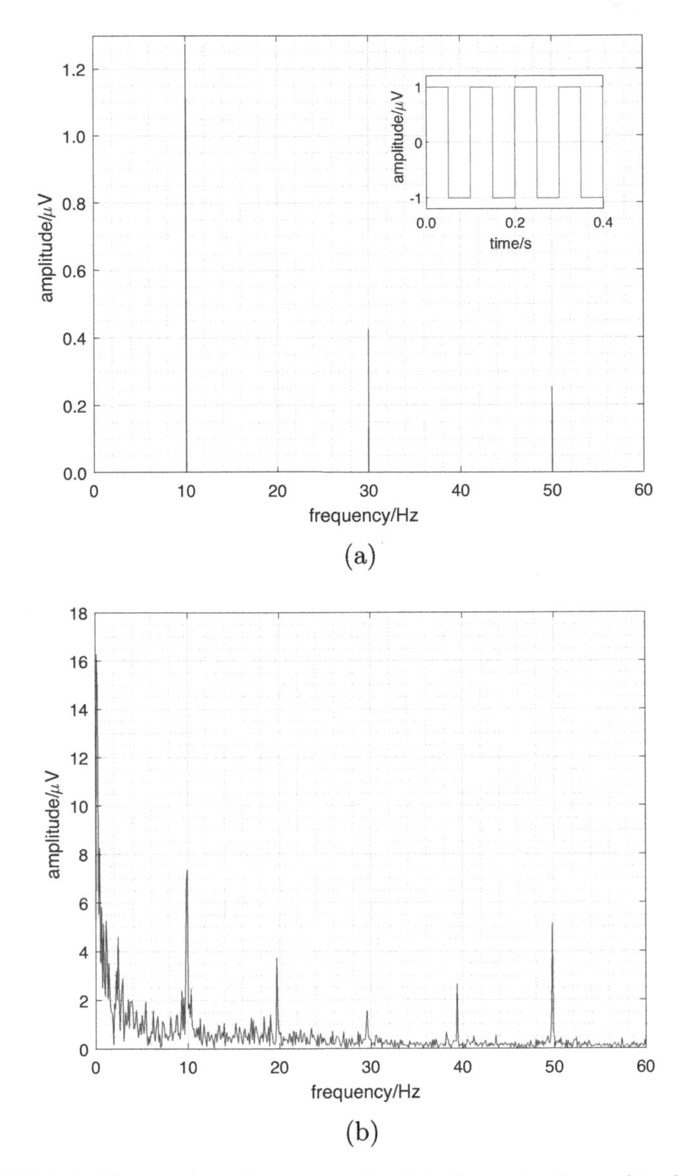

FIGURE 3.2 Comparison between simulated nominal amplitude spectrum of the visual stimuli (a) and measured EEG spectrum corresponding to visual stimulation (b).

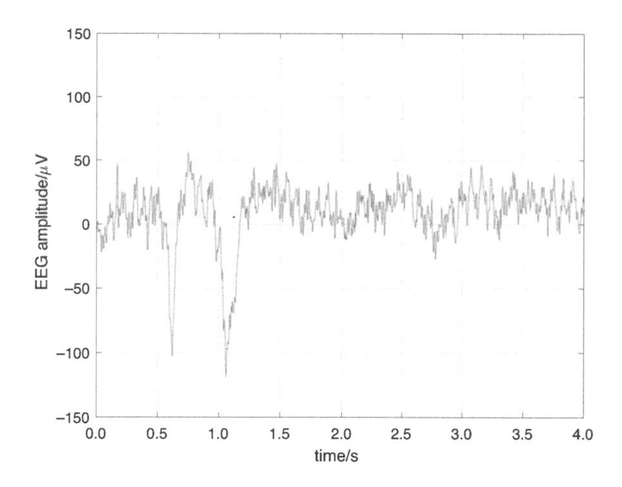

FIGURE 3.3 A typical electroencephalographic signal recorded through dry electrodes during experiments with SSVEP. Artifacts related to eye-blinks are clearly detectable by the pronounced valleys in the recording.

beneficial for reducing the contact impedance. Also recall that, if an Helmate-like EEG cap [62] is used, the electrodes shape is also optimized to overcome the hair and reach the scalp.

Finally, the contact impedance is lowered by applying pressure to the electrodes. The values reported in [126] for the impedance are also compatible with the case of skin abrasion, thus indicating no need to scrub the scalp surface. In this book, tight headbands are illustrated in aiming to apply a sufficient pressure to the electrodes while avoiding discomfort for the user.

The above-mentioned precautions are indeed useful in daily-life EEG applications because they would require gel-free electrodes. In addition, monitoring the electrodes contact impedance during EEG measurement would be desirable. Impedance is typically checked by an active measurement method: a small current is injected and the resulting voltage is measured [127, 128]. Unfortunately, this is not always possible. For example, the commercial EEG setup from Olimex [61] does not include an impedance check circuitry, and this should be purposely added. Given that, the possibility of different quality checks on the measured signal was considered during this book. In [29] it is suggested that the line noise (50 Hz in Europe) is not properly rejected if there is a difference between the contact impedances associated with the reference electrode and the measuring electrodes. In particular, in experiments with the EEG-SMT by Olimex [61], it was observed that the 50 Hz harmonic was emphasized when the two electrodes of a differential channels were applied with different pressures. By noting that one of the electrodes was placed on the forehead (no hair), this eventuality suggested to adjust the electrode in the hairy scalp area, usually by tightening the headband.

Actually, even if this is to be done during the user preparation, the electrodes placement slightly change during the experiments due to their long-term instability. This can be also detected by the appearance of a higher 50 Hz harmonic in the recorded signals, but at least in experiments with SSVEP, such an event rarely compromises the brain signals classification. In accordance with such experimental experiences, it appears that contact impedance constraints are not always tight, since the system functionality remained despite some instabilities. Nonetheless, this can be true for the SSVEP detection, while it was not possible to observe a similar phenomenon in the experiments with motor imagery.

As a further check for signal quality relying on the measured EEG signals, artifacts have to be visually inspected. Indeed, it is generally accepted that a normal EEG is characterized by the absence of identifiable abnormalities. Hence, a statistical definition of a "clean EEG" signal can help to set threshold values so to determine artifact levels in an EEG recording. These thresholds are generally based on amplitude, skewness, and kurtosis of the EEG signal. Threshold-based approaches are commonly used to reject EEG segments, they require to manually define the thresholds, and distinguishing between high and low level of contamination is not trivial.

In such a framework, classifier-based methods have also been proposed to this aim. As an example, some authors chose to divide the quality of the EEG in three classes [129], i.e., low quality, medium quality, and high quality, and then they employed more than a hundred features to classify the EEG quality of the segments. The considered features were generally extracted from EEG signals filtered in different frequency bands. Meanwhile, only three signal quality indicators were used in [130]: the ratio of alpha power to the total EEG power, the variance of each EEG signal, and the power at the line frequency (50 Hz or 60 Hz). These parameters were motivated by the fact that, when the eyes are closed, the power of alpha rhythm becomes dominant and it can occupy a large part of the frequencies of interest.

Moreover, it appears useful to check line noise because in daily-life settings this is the most prevalent noise source. Overall, there is no unambiguous standard for evaluating the quality of EEG signals, and detecting artifacts typically relies on empirical experience. In the case of dry electrodes, this is even more challenging since the recording is more sensitive to artifacts and the acquired signals would be typically discarded by a clinician used to EEG recording with wet electrodes. As a representative example, Fig. 3.3 shows a 4 s-long EEG signal recorded during experiments with SSVEP. The pronounced valleys in the recording correspond to the movement of the electrodes due to eye blinking, and such artifacts would lead to discarding this trial. However, such signal can be easily classified as corresponding to a 12 Hz stimulation and it should be not discarded at all. In this sense, the empirical experience conducted with dry electrodes differs from the one associated with classical recording methodologies.

Apart from artifacts detection and eventual rejection of a whole trial, artifacts could be also removed from a trial. In [131], a comparison between the most common techniques of artifact detection and removal have been presented, e.g., regression, filtering. Regression requires the recording of electrooculographic (eyes) and/or electromyographic (muscles) activity in addition to the brain signals recording, thus this is typically undesirable in enhancing wearability and portability of the system. Analog and digital filtering is instead quite simple and effective, and it only requires that the identifiable artifact-related bands do not overlap with the signal-related band. Finally, *Independent Component Analysis* (*ICA*) is a typical technique for separating signal sources from noise sources in trying to remove the noisy part, but it requires multi-channel acquisition.

Also, a similar technique is based on *Artifact Subspace Reconstruction* (*ASR*) and it also typically involve multiple channels [132]. Hence, in the BCI systems with a single or few channels, filtering or ICA extensions are typically considered. As it will be detailed in the respective chapters, filtering was sufficient in the SSVEP-BCI, while and extension of ICA was investigated for Motor Imagery. However, artifact removal techniques should be better investigated for the few channels cases. As a general consideration, artifact removal did not appear essential in the case of SSVEP, given that precautions were taken in properly measuring the EEG. Instead, evaluating the signal quality in motor imagery was more challenging due to concomitant unexplained phenomena.

3.2.6 Smart Glasses Characterization

In this subsection, the discussion copes with the possibility to provide visual stimuli through Smart Glasses. Especially because of their wearability and portability [133], these devices are increasingly exploited in *Extended Reality* (*XR*), which is typically declined into *Virtual Reality* (*VR*), *Augmented Reality* (*AR*), and *Mixed Reality* (*MR*) . For a wearable SSVEP-BCI, the user can interact with the AR glasses by merely staring at icons appearing on the display. The functionality of wearable XR-BCI systems has already proved by previous research. However, errors and uncertainties of such a stimuli generator were not quantified. Currently, calibrating Smart Glasses regards the accurate measurement of both the device physical position and the user's eyes position for properly rendering virtual objects [134]. Even in dealing with display calibration, pixel-wise calibration is needed for spatial objects positioning [135]. Clearly, generating flickering icons with Smart Glasses is strictly related to the BCI framework, and, even in SSVEP-BCI research, previous works only dealt with the optimal layouts for the stimuli [136]. In this subsection, instead, we focus on the display characterization from the view of flickering icons generation.

Commercially available Smart Glasses exploit different display technologies. A basic distinction is between video see-through and optical see-through

[133]. The former relies on an embedded camera to record the surrounding environment and then display virtual objects overlapped to the recorded video. The latter, instead, uses a semi-transparent display that does not hide the real scene while superimposing virtual objects on it. Although the two technologies have many common features, optical see-through devices are mostly considered in this book because they are better suited for AR applications.

A first example of optical-see-through device is the Microsoft HoloLens [137]. The HoloLens display consists of a set of transparent screens, showing a different image to create a stereoscopic illusion. The displays are planar waveguides: a source transmits image data along the length of the transparent displays, and then the light rays get eventually extracted to reach the user's eyes. To generate a flickering icon, the HoloLens can be programmed by considering that the declared refresh rate is 60 Hz. Therefore, these icons are obtained by switching on and off display pixels.

Another family of Smart Glasses is the Epson Moverio BT200 and the BT-350 [138]. The first has an LCD display with an active matrix of polysilicon thin-film-transistor. The second one, instead, exploits a silicon-based organic LED matrix. Both devices can be programmed in Android with a dedicated graphic library. The screen refresh rates are 60 Hz and 30 Hz, respectively.

In a straightforward approach, the flickering frequencies are obtained by switching pixels on or off after a fixed number of refresh periods. Thus, the achievable flickering frequencies equal the refresh rate divided by an integer number. More sophisticated approaches could be also adopted to achieve more frequencies, but this would go beyond the scope of this book. In SSVEP applications, higher refresh rates are usually preferred to enable more flickering frequencies. This was a main reason for employing the BT200 in the BCI implementation, and, as a consequence, the stimuli characterization primarily concerns this device. However, further information about the flickering generated with other display technologies is also reported in the following.

4

SSVEP-Based Instrumentation

This chapter deals with the realization of a SSVEP -BCI system. In particular, design, prototype and performance of the system are analyzed in Section 4.1, Section 4.2, and Section 4.3, respectively.

4.1 DESIGN

The architecture of the AR-BCI system is shown in Fig. 4.1. Its building blocks are:

- AR glasses employed as a visual stimuli generator;
- EEG transducer acquiring the brain signals;
- computing unit to process brain signals and generate a control command again inside the AR glasses.

The system allows the user to control AR glasses with his/her brain activity instead of a touch pad or any other controller provided by default. As already discussed, SSVEP-based BCIs require visual stimulation with flickering icons. The display of AR glasses can provide this stimulation while being wearable, in contrast with mostly adopted solutions based on LEDs. The flickering icons on AR glasses display correspond to choices for the user during the usage of an application, and thus they substitute a more traditional interface based for instance on a touch-pad or gestures. The potentials evoked by the flickering are then measured with an electroencephalography. Note that the input interface for the inspection system is made of either the EEG transducer, the computing unit for processing, and also the visual stimulation. Moreover, the computing unit and output interface of the AR glasses manage the BCI application also in terms of the visual information resulting from the user's control. Another picture of this SSVEP-BCI system was already presented in Fig. 3.1, while the system worn by a user will be shown next. The implementation of each system block is instead discussed hereafter.

In signal acquisition, different aspects are taken into account. Firstly, brain activity has to be measured with a single-channel non-invasive EEG, due to

DOI: 10.1201/9781003263876-4

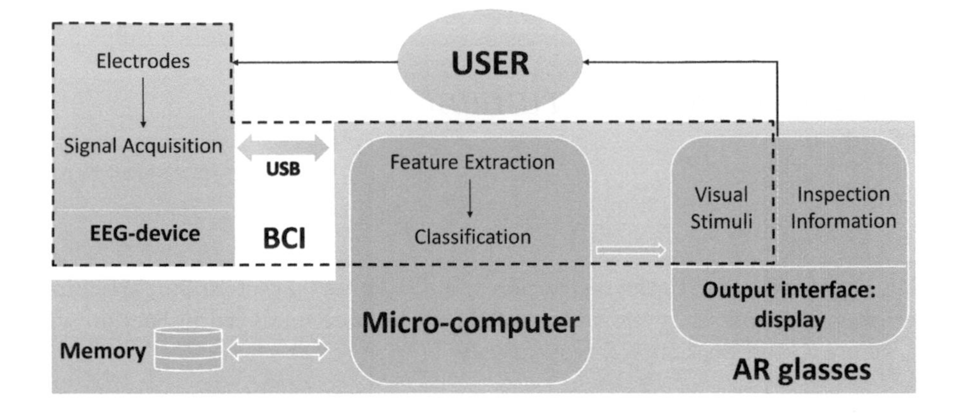

FIGURE 4.1 Architecture of the AR-BCI system based on steady-state visually evoked potentials [31].

the practical reasons, notably user friendliness and low cost. Secondly, dry electrodes are chosen because electrolytic gels would pose some issues in long-term usage, for instance the need to repeatedly apply the gel that dries out. Dry electrodes are easier to be worn and they might still provide proper EEG measures. On the other hand, avoiding conductive gels poses a severe issue on electrode-skin contact: if the contact impedance is too poor, the EEG activity is not present in the recorded signals. To this aim, tight bands have to be used to ensure a good electrode-skin contact.

Finally, EEG signals must be processed in order to recognize the user's intention. The final aim is to assign a class to each signal recorded for a certain time, and that class should be associated with the user's intention, i.e., the control command for the application. Two main parameters will be used to characterize the performance of the system: the classification accuracy, namely the success rate in classifying the brain signals, and the latency for the system to react, consisting of acquisition time plus elaboration time required to make a choice. One may guess that there will be a trade-off between classification accuracy, and latency. The details of the system are discussed in the next subsections.

4.2 PROTOTYPE

This section provides details of the hardware solutions, i.e., Smart Glasses and acquisition system, and implemented processing strategies.

4.2.1 Augmented Reality Glasses

As a reactive paradigm, the SSVEP-BCI requires flickering icons on the AR glasses display for visual stimulation. In particular, the specific VEP exploited in this book regards the generation of an oscillation at the same frequency of the flickering stimulus. Hence, when staring at a specific icon, the user's intention is easily retrieved from the frequency domain. Each icon flickers at a specific frequency, and these frequencies must be carefully chosen. Also note that an amplitude modulation could not be considered to encode the information of each possible choice. The reason for that lies in the inter-subject and intra-subject variability of SSVEP signals, but also because of the poor amplitude accuracy of the off-the-shelf components adopted in the system implementation. A phase modulation, instead, would require a higher stability of the flickering stimuli, but the AR glasses characterization suggested that this is not the case for a typical commercial device. In conclusion, these techniques would require further investigations, beyond the scope of this book.

In its simplest implementation, the SSVEP-based BCI exploits only two icons. A custom Android application is developed so that these icons could be placed on the opposite corners of the AR glasses display. Moreover, referring to literature studies, white squared icons were implemented to optimize the stimulation [139]. The flicker frequencies were chosen in the *alpha* band (8 Hz to 13 Hz) according to a study indicating this range as the one associated with maximum amplitude of elicited oscillations [110]. Specifically, the nominal flickering frequencies were set at 10.0 Hz and 12.0 Hz. By exploiting the results of Section 3.2.6, the related uncertainty was estimated to be in the order of 0.1 Hz. The Android "open graphic deepl librar" (*openGL*) was used in implementing the icons so that the GPU could manage the flickering. Indeed, avoiding the use of CPU in this thread aims to maximize flickering stability, which could be affected by operating system interruptions.

To make a choice through a flickering icon, the user has to stare at (or simply focus on) the desired icon for some seconds. This interval corresponds to the EEG acquisition time. As a first implementation, such a time was fixed a-priori, while a variable time window can be also used by exploiting a stopping criterion for the flickering. Such a possibility will be recalled later on as a possible variant of the basic system functionality. The Epson Moverio BT-200 [138] is used as a communication and control interface, namely the AR device for both visual stimulation and visual interface with the application. The Android application running on such glasses acquires the EEG signals in parallel to visual stimulation, while the processing of those signals is conducted at the end of the acquisition window. An attentive reader could guess that the processing should be instead conducted in real-time when the acquisition time is not fixed a-priori. Acquisition and processing are detailed next. At the end of those steps, the user's intention is retrieved, short of misclassification errors. The control command triggers an action in the application, e.g., one could request to read data from a sensor. The chosen AR device is capable of communicating with external devices through Bluetooth or WiFi. Hence, the information of interest for the user can be requested without using hands thanks to the EEG transduction, and wireless communication and control is possible.

4.2.2 Single-Channel Electroencephalography

In the architecture presented in Fig. 4.1, the input interface of the system is a non-invasive single-channel BCI based on an EEG device. Only two dry electrodes are employed for a differential acquisition of brain signals. This allows one to have an utmost wearable system. The electrodes are placed on the scalp, according to the 10-20 system [30] at the points *Fpz* and *Oz*, as it was already highlighted in black in Fig. 1.5. This choice was done by considering that *Oz* is located at the occipital region of the brain, the one associated with visual activity, while *Fpz* is at the frontal region, where it is reasonable to assume limited visual activity. Therefore, a differential acquisition aims to emphasize the only visual activity by subtracting part of the ongoing brain activity. To mitigate the contact impedance issues, silver pins were soldered on the occipital electrode, which needs to overcome the hair to reach the scalp. Furthermore, the use of active dry electrodes eases impedance matching by buffering recorded signals on the electrode itself. In particular, the active electrodes include a circuitry based on an operational amplifier for impedance matching. A third passive electrode, i.e., without active buffering, acts as a ground for the measurement. This ground electrode is usually placed on the forehead, the ear, or even on a wrist or a leg [28]. In this book, two positions are exploited: the left wrist and the left ear. No meaningful difference was found between these two placements, at least in terms of system functionality. Meanwhile, placing the passive electrode with a clip on the ear guarantees better mechanical stability of the passive electrode, which surely results in a more reproducible measurement.

The Olimex EEG-SMT [61] with dry electrodes is chosen for the implementation of the EEG transducer. One of the main reasons for this choice was the very-low cost if compared to more classical EEG amplifiers. It is worth remarking that the discussion of Section 2.4 assessed the metrological properties of this device, which proved adequate for wearable BCI applications. Nonetheless, the skin-electrode impedance could not be studied with a mere electrical characterization of the device. This had to be proven by experimenting with human users, and such contact impedance issues resulted in the most challenging aspect in EEG acquisition. In the prototyping phase, a good electrode contact is ensured with a tight headband for the active electrodes on the user's scalp.

The electrical brain activity measured with the electrodes is transmitted to the EEG transducer board. The two active electrodes are connected to the differential input of an instrumentation amplifier at channel 1 (CH1). A twin channel (CH2) is available in the Olimex EEG-SMT board, but it is not used in the SSVEP-BCI, so it has to be internally short-circuited. Meanwhile, the passive electrode provides feedback for instrumentation amplifiers to reduce common mode noise. In the Olimex EEG-SMT, this is connected to the input DRL. Signal conditioning is done with several stages providing amplification and/or filtering. The overall gain from the signal pick-up to an ADC was set to 6427 V/V thanks to the second amplifying stage, which grants an adjustable gain. The input signal is analogically filtered with a 3rd order Butterworth filter, whose nominal pass-band is 0.16 Hz to 59 Hz.

At the end of the amplification and filtering chain, an ADC with 10-bit resolution provides the conversion to a digital signal with a nominal sampling frequency equal to 256.0 Sa/s.

The digital signal can thus be transferred to the computing unit of the AR device to be processed. Note that the Olimex EEG-SMT continuously acquires and sends the digitized signals over UART connection (with an USB cable for instance), so that the Android application can save only the EEG epochs of interest.

4.2.3 Data Processing

In the previous discussion, it was largely anticipated that the frequency domain enables an easy and intuitive detection of the SSVEP oscillations. Another possibility consists in processing the signal in the time domain. As will be discussed later, the analysis in the frequency domain, apparently most natural for the detection of the SSVEP oscillations, has the disadvantage of having a higher latency. The analysis in the time domain, on the contrary, has a lower latency but is also characterized by a lower accuracy. Therefore, two possible strategies of signal processing are described in detail below: in the time and in the frequency domains.

Frequency domain analysis

The reason for considering a power spectral density analysis is to guarantee low computational burden in implementing a wearable system. Moreover, the good knowledge on the SSVEP phenomena makes it useless to adopt computationally challenging approaches, such as Deep Neural Networks, which would also require much EEG data. To present the processing steps of interest, it is useful to distinguish the *feature extraction* and *classification* steps.

In *feature extraction*, the digitized signal is pre-processed with a digital pass-band filter based on a *Finite Impulse Response* (*FIR*) paradigm. After some preliminary trials, an optimal band is identified, the range 6 Hz to 28 Hz, and order 100 are set to have at least 50 dB of attenuation in the stop band. The pass band is also chosen by considering the need to reduce eye-blinking and muscle artifacts [32]. Indeed, linear filtering could be used to reduce noise introduced by artifacts, while methods like regression would need for a higher number of electrodes. After that, zero-padding is applied prior to executing a

Fast Fourier Transform (*FFT*). Zeros are thus added at the end of the signal samples to reach the nearest power of 2, so to speed up the FFT execution and possibly provide better frequency domain resolution. Note that zero-padding can fictitiously enhance resolution only because of interpolation, while the actual spectral resolution depends on the length of the EEG epoch, i.e., the acquisition time window. This step makes clear the trade-off between system latency and classification accuracy: a longer acquisition time enables better frequency resolution and hence better SSVEP detection, since frequency resolution is the inverse of time window. Conversely, lower latencies are desirable to avoid long selection times for the icons, but this limits the resolution in detecting SSVEP peaks. Furthermore, the Hamming windowing is applied to reduce spectral leakage, while slightly losing in spectral resolution.

Once the signal under analysis is properly represented in the frequency domain, Power Spectral Density (PSD) analysis is carried out on the discrete amplitude spectrum. For each frequency f_i, the corresponding PSD is calculated as the sum of the squared amplitudes associated to the k_i-th bin (the interval around the center frequency, whose width is the spectral resolution) and some nearest bins:

$$P(f_i) = \frac{1}{2k+1} \sum_{n=k_i-k}^{k_i+k} A^2(n), \tag{4.1}$$

where k_i is the bin associated with the frequency f_i, k is the number of bins considered on the right and on the left of the bin k_i, and $A(n)$ the amplitude associated to the n-th bin. The bin number k_i is retrieved from f_i by means of the spectral resolution Δf of the FFT, at least in principle, since $f_i = k_i \Delta f$. The frequency f_i can be equal to the nominal stimulation frequency or a higher harmonic of it (an integer multiple). However, the SSVEP peak could not be exactly located at f_i, and hence there is a preliminary step for finding the maximum in the neighborhood of k_i is found. The considered neighborhood is 0.4 Hz. For the sake of the clarity, this bin is referred to as k_i'. Extracting the PSD leads to a new representation of an EEG signal in a features domain. Two possible representations are thus investigated: the first considers the PSD at 10 Hz and 12 Hz, while the second also considers the respective second harmonics, i.e., 20 Hz and 24 Hz. In the former case, each signal is represented as a point in a plane, while in the latter one it is represented as a point in the 4D space. Typically, higher-order harmonics are not considered because of the poorer signal to noise ratio, but they could also contribute to signal classification in some cases.

As far as the *classification* is concerned, in the features domain representation, a hyperplane is adopted to separate the two classes of signals. Specifically, a SVM was used by investigating the linear and the Gaussian kernel. After some preliminary trials on EEG signals, the linear kernel resulted in the optimal solution. Hence, the actually adopted classifier is a *Support Vector Classifier* (*SVC*) [33]. The analysis of EEG signals and the training of the classifier is done in Matlab. The training must be conducted with labeled EEG signals, i.e., signals for which one knows the belonging class (in the present case, 10 Hz stimulation or 12 Hz stimulation). Notably, Matlab is really useful in deriving the SVC hyperplane parameter that can be then used to classify new unlabeled signals. However, both the features extraction and the classification modules were implemented in Java. This step was necessary in building a wearable system based on Android. For the classification step, implementing the SVC simply consists in solving an inequality considering the hyperplane parameters and the dot associated with the signal to classify.

Time domain analysis

As aforementioned, another possible strategy consists in analyzing the signal in time domain. In particular, correlation-based algorithm are implemented. A correlation-based algorithm for frequency evoked by flickering stimulus recognition is then described.

The first step of the algorithm consists in applying a pass-band FIR filter (5 Hz to Hz) to the signal divided in time windows of duration T (Fig. 4.2). The second step consists in (i) comparing the filtered signal D_F with two

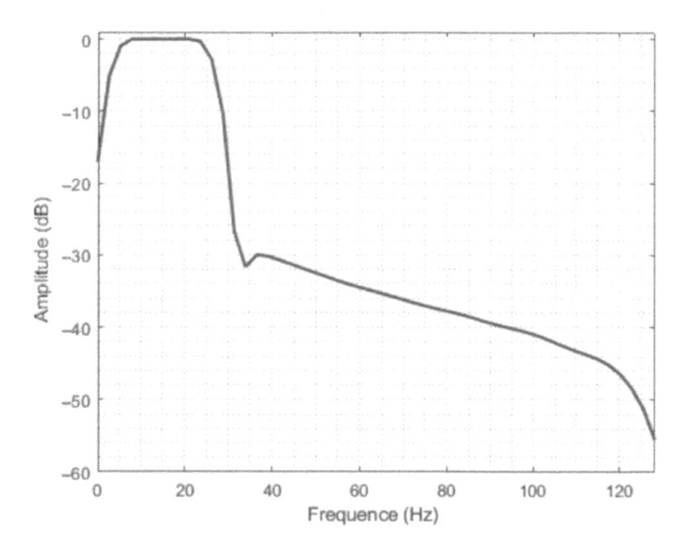

FIGURE 4.2 FIR Filter Amplitude frequency response [140].

sinusoidal signals, Φ_1, and Φ_2-whose frequency corresponds to flicker stimulus frequencies and - whose phase ϕ is variable) and (ii) in finding the maximum Pearson correlation coefficients (ρ_1 and ρ_2):

$$\rho_1 = \max_{\phi \in [0,2\pi]} \frac{cov(D_f, \Phi_1(\phi))}{\sigma_{D_f} \sigma_{\Phi_1(\phi)}} \tag{4.2}$$

$$\rho_2 = \max_{\phi \in [0,2\pi]} \frac{cov(D_f, \Phi_2(\phi))}{\sigma_{D_f} \sigma_{\Phi_2(\phi)}} \tag{4.3}$$

where σ_D is D_f standard deviation, and σ_Φ the sinewave standard deviation. Then, F1 and F2 features are:

$$F1 = max(\rho_1, \rho_2) \tag{4.4}$$

$$F2 = \frac{max(\rho_1, \rho_2) - min(\rho_1, \rho_2)}{min(\rho_1, \rho_2)} \tag{4.5}$$

Then, the following condition is checked:

$$F1 > T1 \quad \cap \quad F2 > T2 \tag{4.6}$$

where *T1* and *T2* are the thresholds. In case this condition is not met, the signal is considered as detected, otherwise, a new signal, overlapping with the previous one by *T/2*, is analyzed.

In this SSVEP algorithm, the latency is the time necessary to detect one of the two frequencies.

Finally, a threshold algorithm is used to distinguish voluntary from involuntary eye blinks (Fig. 4.3). In fact, voluntary eye blinking is characterized by higher peaks than the involuntary one [141].

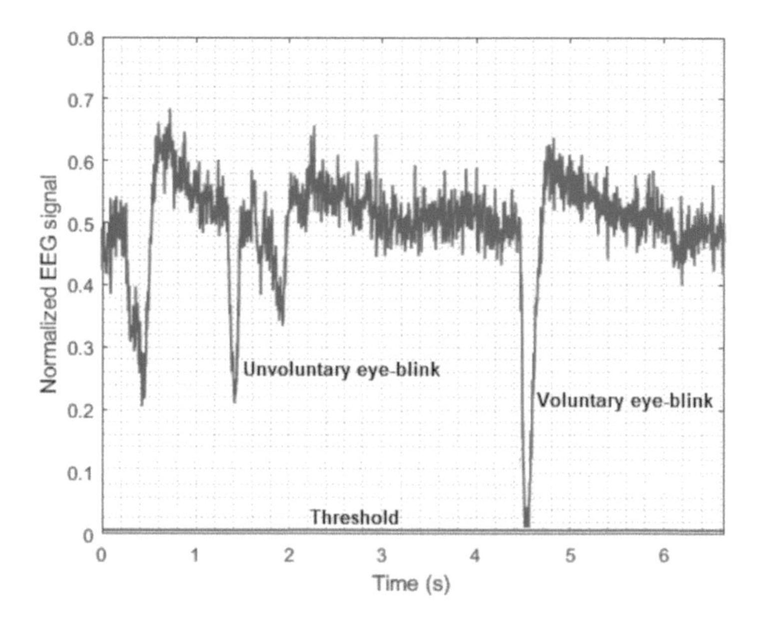

FIGURE 4.3 Detection of voluntary and involuntary eye blinks [142].

4.3 PERFORMANCE

After the design of the BCI system based on SSVEP, and its prototyping by means of the commercially available components mentioned above, validation is necessary. In the following, some details about the experimental validation are reported with specific regard to the setup, the experimental conditions, the assumptions, and the involved subjects. Then, the results in terms of classification accuracy are discussed, and a particular focus is given to the trade-off between the stimulation time (greatly affecting the system latency) and this classification accuracy.

The performance of the system is analyzed with reference to processing before in the time and, then, in the frequency domain.

4.3.1 Frequency Domain

Experimental campaign

Twenty subjects take part in the experiments, 13 males and 7 females, with age between 22 and 47 years old. Experiments are conducted in a laboratory with closed blinds, so that luminance could be controlled with neon lights. Illumination is hence monitored during the experiments and it results in the range 95 lx to 99 lx. All the electrical and electronic instruments present in the room are switched off to avoid interferences with the EEG. A laptop is used to acquire from the EEG transducer during the first experimental phase. However, it is unplugged from the mains supply during signal acquisitions. Another electronic device in the room is the battery-supplied AR device.

Although the final aim is to build a portable system, each subject under test is asked to sit on a comfortable chair with armrests and to limit unnecessary movements. Once seated, the user could wear the system. The AR glasses has to be worn first, and then a tight headband can be used to fix the electrodes at the occipital and frontal region of the scalp. Also note that in these experiments a tight armband is used to fix the ground passive electrode on the left wrist of the user. In connecting the electrodes, the same sequence is adopted for every user while checking in real-time the acquired signal:

1. the passive electrode is connected to the input Drive Right Leg of the EEG transducer and applied on the subject's left wrist with the armband; the acquired signal has to be null due to the absence of signal at CH1 (or CH2);
2. the first active electrode (without silver pins) is connected to the negative terminal of CH1 and placed on the scalp at *Fpz* (frontal region) with the help of the headband;
3. after a transient (lasting a few seconds), the acquired signal has to return to zero again, since CH1 is still an open circuit at this stage;

indeed, the internal circuitry reaches a stationary condition with a null output;

4. lastly, the second active electrode (with silver pins) is connected to the positive terminal of CH1 and placed on the scalp at *Oz* (occipital region) with the help of the headband; after another transient of a few seconds, a stationary condition is reached in which the mean signal amplitude is null;

The EEG acquisition board is connected to the laptop to acquire signals. At the beginning of each test, the EEG signal amplitudes has to be checked: it was determined empirically that proper electrodes placement is associate with signal oscillations with a peak-to-peak amplitude below 100 μV. However, because of dry electrodes, some artifacts can be occasionally present, notably the ones associated with subject's eyes blinking. Unfortunately those artifacts cannot be avoided, and in case of badly placed electrodes they led to amplifier saturation. At least 1 s is typically needed to recover from saturation, and hence such a condition has to be avoided. On the other hand, when an artifact related to eye-blink is present without any saturation, signal disruption appears not meaningful, as already shown in Section 3.2.4. A typical signal measured with the worn EEG transducer is represented in Fig. 4.4. The artifact related to eye-blinks is revealed with the presence of negative valleys. Note that this phenomenon is less present in EEG systems employing wet electrodes. Given that, the presence of such valleys is reasonably led back to electrodes movement, while it seems unreasonable to read those valleys as neural activity associated with eye movements. As a further proof of that, similar artifacts appear on the recorded signal when moving the electrodes in other ways. It is clear that such a system cannot be used in everyday life without a robust artifact removal block, which is beyond the scope of this book.

For each subject, first 24 trials with a single flickering icon are carried out, and then other 24 trials with two flickering icons are conducted. In the first set of trials, the flickering frequency is randomly chosen between 10 Hz and 12 Hz in order to avoid user biases. According to the design, the 10 Hz flickering icon appears on the bottom-right corner of the AR glasses display, while the 12 Hz on the up-left corner. In each trial, the brain signal is acquired for 10.0 s, and a few seconds passes between consecutive trials. Note that 10.0 s is chosen as an upper limit for user stimulation, which can be already unacceptable in several practical applications. However, smaller time windows can be analyzed too by properly cutting the recorded signals and retrieving smaller epochs. The set of trials with two simultaneously flickering icons follows the same principles, but the user has to randomly choose the icon to stare at. In both cases, the only constraint is that the subject has to stare at both icons for 12 times each. At the end of a trial the subject has to declare the choice he/she made, so as to have a class label for each trial. In doing this, there is no guarantee that the user was effectively able to focus the attention on the declared icon

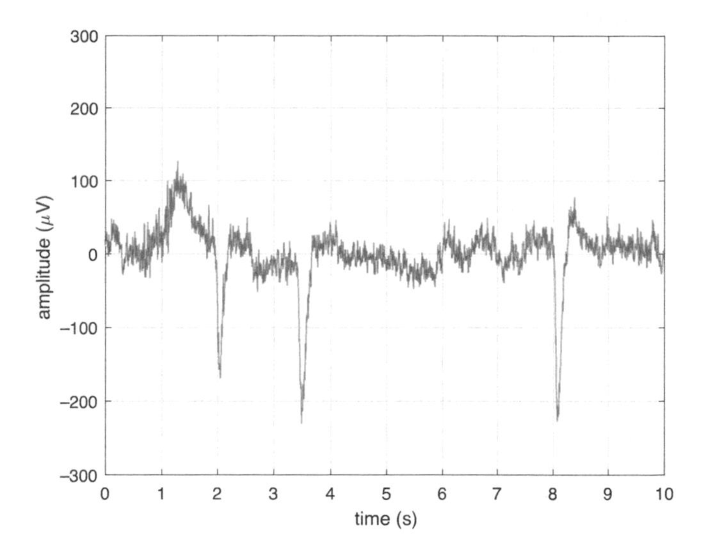

FIGURE 4.4 An example of a signal measured with the Olimex EEG-SMT after placing all the dry electrodes for SSVEP detection. Note that some artifacts related to eye-blinks are present.

for the whole acquisition time. Moreover, the phase of the flickering icon is not synchronized with the acquisition starting. For this reason, the phase information of the signals is not exploitable for these EEG data.

The data acquired in these conditions are used to train and test the processing of SSVEP signals. In this phase, no results are given on the AR glasses display and the data analysis is conducted offline. This allows to validate the system design and also to make adjustments where needed. Then, in a later phase, experiments are conducted with the signal processing directly implemented on the AR glasses computing unit, thus avoiding the usage of the laptop with Matlab in favor of the final system implementation. In this way, the BCI system is really wearable and portable and it could be exploited in different applications. These applications are described at the end of this chapter.

Classification results

The EEG data acquired from each subject during the visual stimulation are analyzed as described in subsection 4.2.3, namely by representing each signal in terms of PSD at the stimulation frequencies or their multiple harmonics. Different aspects can be considered in discussing the classification of such signals.

First, it is useful to compare the class separability between EEG signals corresponding to 10 Hz stimulation and 12 Hz stimulation. Figure 4.5 show the signals of all subjects in the PSD features domain. In particular, each signal is a point in the 2D plane identified with the respective PSD at 10 Hz (x-axis) and 12 Hz (y-axis). A 10 s-long stimulation is currently considered, and the case of a single stimulus is compared with the case of two simultaneous stimuli. The signals corresponding to the two different stimulation frequencies are distinguished by their color. As a general trend, the two classes are quite separated in the features domain.

However, there is an overlap between the two classes and some of the signals fall into the "wrong region": for instance, one of the signals labeled as *class* 12 Hz in Fig. 4.5(a) is very far from other signals of the same class and much closer to the *class* 10 Hz. Note that better class separability could be expected for the "1 stimulus", while in the "2 stimuli" case one could foresee an interference of the other stimulus while the user is trying to stare at one of them. Nonetheless, qualitatively speaking, there is no such an evidence in comparing Fig. 4.5(a) and 4.5(b). On the contrary, the classes may seem better separated in the "2 stimuli" case. This will be better assessed in a while by quantifying class separability with classification accuracy.

It is also interesting to consider a shorter stimulation time: this can be done in post-processing by considering fewer samples of the acquired signals. In doing so, a 2 s-long stimulation time is taken into account. Surely, a shorter stimulation is desirable to speed up communication and/or control with the SSVEP-based BCI, and the trade-off between system latency and classification accuracy will be extensively discussed in the following subsection. Either way, Fig. 4.6 anticipate that SSVEP classification is less accurate in such a case because, as expected, classes are less separable. It would be also interesting to distinguish between different subjects, so to graphically highlight if there exist "good" and "bad" subjects. Nevertheless, analyzing 20 subjects with scatter plots is not manageable and assessing the classification accuracies was needed before further considerations. Finally, as mentioned above, further PSD features can be considered, but a geometrical representation is challenging for a 3D space and impossible for a higher-dimensional space. Also in this aspect, considering the classification accuracy is essential.

Several metrics could be used to quantify the separation between classes. As an example, for each class a center could be identified and the Euclidean distance between these centers can be assumed as a measure of separability. However, hereafter the classification accuracy is considered since it is well related to the BCI system performance. As discussed previously, this metric is obtained as the ratio between the number of correctly classified signals (according to their known class label) over the total number of signals to classify. Therefore, such a metric is directly related to the success rate of the SSVEP detected by the BCI system. It is even clear that knowing the true labels for all the signals is crucial in assessing the classification accuracy. In the present case, the subject had to declare at the end of each trial the chosen stimulus, so that the probability of a wrongly assigned label is non-null. This is also true for the "1 stimulus" case, where the user could manually select

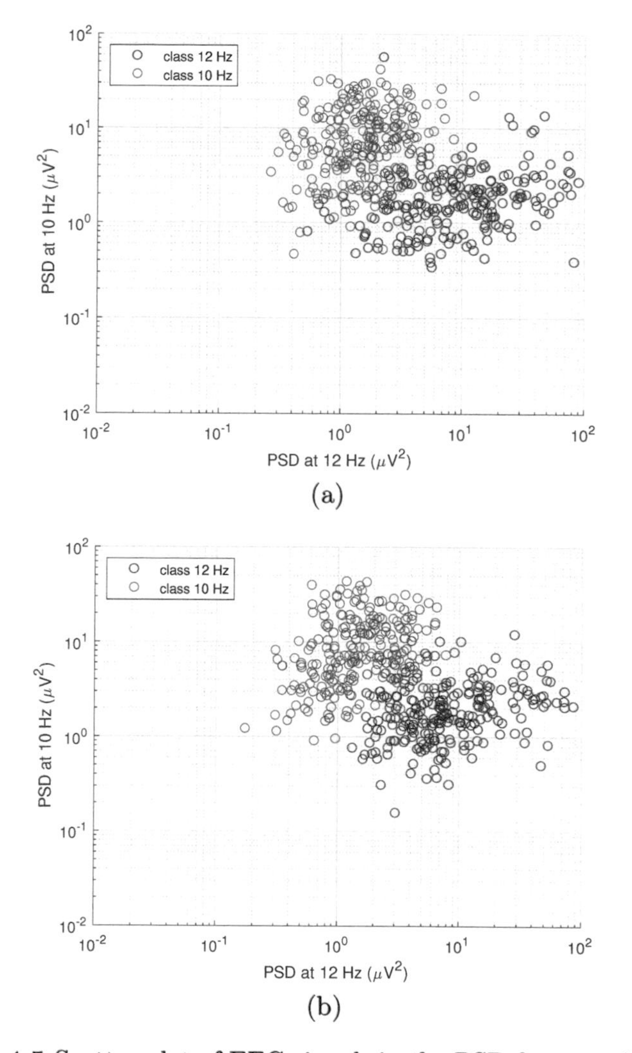

FIGURE 4.5 Scatter plot of EEG signals in the PSD features domain, associated with 10 s-long stimulation and for a different number of simultaneous stimuli. Both axes are in logarithmic scale. (a) 1 stimulus and (b) 2 stimuli.

the unique stimulus appearing for each trial and declare the choice at the end of it. Although these errors are possible, it was reasonably assumed that the number of wrong labels is actually negligible. Therefore, the main causes for a classification accuracy lower than 100 % are: poor concentration during some experimental trials, low SSVEP activity, EEG measurement errors, such as temporary electrode disconnection, and eye-blink artifacts, or the classification model itself.

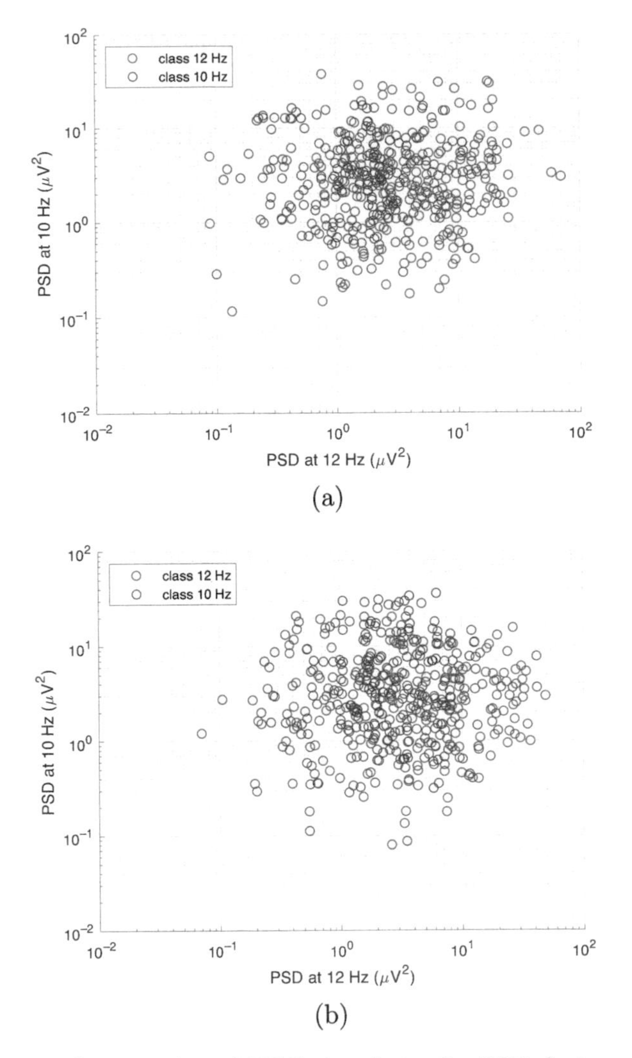

FIGURE 4.6 Scatter plot of EEG signals in the PSD features domain, associated with 2 s-long stimulation and for a different number of simultaneous stimuli. Both axes are in logarithmic scale. "1 stimulus" (a) and "2 stimuli" (b).

When identifying the classification model, some trials are needed together with their true labels to tune model parameters: in Machine Learning, these are referred to as the *training set*. Then, a *test set* is needed to validate the identified model. In the current analysis, cross-validation [143] was used to iteratively split the available data into training and test, so to achieve for each split a classification accuracy on the test set. The mean accuracy between different splits gives an estimate of the model accuracy on unseen data.

In particular, a 4-folds cross-validation is considered, thus splitting the data in 75 % for training and 25 % for testing four times. Together with the mean classification accuracy, the standard deviation can be obtained too, so to quantify the precision of this mean. Table 4.1 report the cross-validation accuracy (with associated standard deviation) for the 20 subjects considered as a whole and one-by-one, for either the "1 stimulus" and "2 stimuli" cases, as well as for a 10 s-long and 2 s-long visual stimulation. Note that the subject "S1" had no trials for the "1 stimulus" case due to measurement issues, so there is no accuracy to be calculated. It can be seen that, as anticipated, a longer stimulation outperforms a shorter one, while there is no meaningful difference between the "1 stimulus" and the "2 stimuli" case. Hence, in this setup, the non-gazed stimulus is not generally disturbing the focus on the other one. It is worth remarking that in both the cases the classification happens among two classes, while increasing the number of total stimuli would indeed reduce the classification performance.

The results in Tab. 4.1 show that, for the 10 s-long stimulation, eight subjects out of 20 reach 100 % in accuracy, while only one subject reaches 100 % in 2 s for the "2 stimuli" case. The worst accuracies are about 65 % to 70 % in 10 s, while for 2 s they drop down to about 50 %. Also note that the row "all" corresponds to the case the data from all subjects is considered as a whole. Interestingly, the associated accuracies are very close to the mean accuracies, thus indicating that there is no need to train the algorithm subject by subject. This aspect is indeed very important in wanting to build a BCI system for daily-life applications, because it indicates that a new subject should not lose time in a training session. Instead, the BCI algorithm can be trained on data from previous subjects.

These results can be even enhanced by considering two more features: the PSD at 20 Hz and 24 Hz. In doing that, results similar to the ones of Tab. 4.1 can be obtained. It is then interesting to compare the mean accuracies in the four cases, as reported in Tab. 4.2. These accuracies increase (or remain constant in one case), and at least in the 10 s-long stimulation case the associated standard deviation diminishes, thus indicating less variation among classification performance for different subjects. A statistical test (matched paired t-test [144]) reveals that these increases are not statistically significant. However, the improvement is substantial at least for the 10-s case since the minimum accuracy rises from 65 %–70 % to 77 %–85 %. Further considerations about the 2-s case are instead reported in the following discussion about the latency/accuracy trade-off. Finally, note that the accuracy is recalculated for the row "all" in this 4D SVM case, again one finds that they are really close to the mean accuracies already reported in Tab. 4.2.

TABLE 4.1 Classification performance of SSVEP-related EEG signals. For each subject, the "1 stimulus" case is compared with the "2 stimuli" case, and the results of a 10 s-long stimulation are compared with a 2 s-long one. Performance is assessed with cross-validation accuracy and its associated standard deviation over 4-folds. The mean accuracy among all subjects is reported too, as well as the accuracy obtained by considering all subjects together (row "all"). The SVM classifier considers two PSD features.

	10 s-long stimulation				**2 s-long stimulation**			
	1 stimulus		**2 stimuli**		**1 stimulus**		**2 stimuli**	
Subject	**acc%**	**std%**	**acc%**	**std%**	**acc%**	**std%**	**acc%**	**std%**
all	91.2	3.1	94.2	1.8	74.8	4.8	77.5	1.2
S01	100.0	0.0	100.0	0.0	75	12	100.0	0.0
S02	100.0	0.0	100.0	0.0	80	15	98.8	4.4
S03	100.0	0.0	100.0	0.0	95.8	7.3	92	10
S04	100.0	0.0	100.0	0.0	81	15	92	10
S05	100.0	0.0	100.0	0.0	78	16	88	12
S06	100.0	0.0	100.0	0.0	94.6	8.8	87	15
S07	100.0	0.0	100.0	0.0	67	19	87	12
S08	91.7	8.4	100.0	0.0	58	15	64	17
S09	88	10	97.9	7.7	72	16	57	12
S10			96.7	6.8			88.8	8.8
S11	80	15	96	10	68	17	64	17
S12	97.1	6.4	95.8	7.3	74	13	78	13
S13	81	14	95.4	8.4	63	15	83	12
S14	75	15	95.4	7.5	48	18	53	18
S15	100.0	0.0	94	11	80	15	80	12
S16	90	12	92.9	8.3	70	16	61	14
S17	97.9	5.6	91	11	78	14	78	14
S18	80	14	89.6	9.8	80	15	61	16
S19	65	12	81	17	51	17	52	15
S20	88	11	71	15	80	16	49	16
MEAN	91		94.9		73		76	
STD	11		7.4		12		16	

Latency versus accuracy

In a SSVEP-based system, performance is typically quantified by simultaneously considering classification accuracy and the stimulation time needed to reach it. The reason is that, as a general trend, the longer the stimulation time, the better the SSVEP oscillation can be detected. Indeed, a longer stimulation allows to improve the signal-to-noise ratio when considering SSVEP oscillations with respect to the ongoing EEG activity. Nevertheless, a too long stimulation tires the user out, and it would be deleterious for classification accuracy. It is also clear that a short stimulation is desirable to speed up the

TABLE 4.2 Comparison of classification performance for a SVM classifier considering two PSD features (2D SVM) and one considering four PSD features (4D SVM) of SSVEP-related EEG data. The mean cross-validation accuracies and their associated standard deviations are reported for the "1 stimulus" and the "2 stimuli" case, and a 10 s-long stimulation is compared with a 2 s-long one.

	10 s-long stimulation				**2 s-long stimulation**			
	1 stimulus		**2 stimuli**		**1 stimulus**		**2 stimuli**	
Classifier	**MEAN%**	**STD%**	**MEAN%**	**STD%**	**MEAN%**	**STD%**	**MEAN%**	**STD%**
2D SVM	91	11	94.9	7.4	73	12	76	16
4D SVM	94.4	7.1	97.2	4.3	75	13	76	15

system. In this context, the system latency corresponds to the time needed to acquire and classify the SSVEP oscillation corresponding to a single icon. If gazing a flickering icon corresponds to selecting it, the shorter the stimulation is and the more commands can be sent in a unit time to the BCI application.

In a practical system, system latency and classification accuracy must be balanced to obtain optimal performance, which means a proper success rate in recognizing the user's intention without taking too much time. In order to take into account both the aspects simultaneously, a useful metric is the ITR [23, 145]:

$$ITR = \frac{1}{L}\left[\log_2 N + A\log_2(A) + (1-A)\log_2\left(\frac{1-A}{N-1}\right)\right], \tag{4.7}$$

where A is the classification accuracy, L is the system latency, and N is the number of possible choices ($N = 2$ in the present case). This quantity is usually expressed in bit/min.

As said above, the system latency should be calculated by summing stimulation time and EEG processing time. In the present setup, while stimulation is at least 1 s-long, processing requires less than 50 ms once the algorithm is already trained. Therefore, it is reasonable to consider system latency practically identical to stimulation time.

In doing this, Fig. 4.7 represents the trade-off under discussion. The two curves represent the cases "1 stimulus" (blue) and "2 stimuli" (red). However, it was already highlighted that there is no statistically relevant difference among them. Hence, the two were mostly represented to have some clues about short-term repeatability of such measures. It can be seen that the two sets of measures are compatible, and that the mean classification accuracy generally increases with stimulation time. The trade-off can be better investigated by first considering the median accuracy instead of the mean. The reason relies on the fact that the SSVEP detection performances are diversified among subjects, and the sample mean could be affected by a single poorly performing subject. Along with the median, it seems useful to represent the *Interquartile Range* (*IQR*) as a statistically robust estimate of dispersion. Such quantities are represented in Fig. 4.8 by considering the IQR between the 75th and the

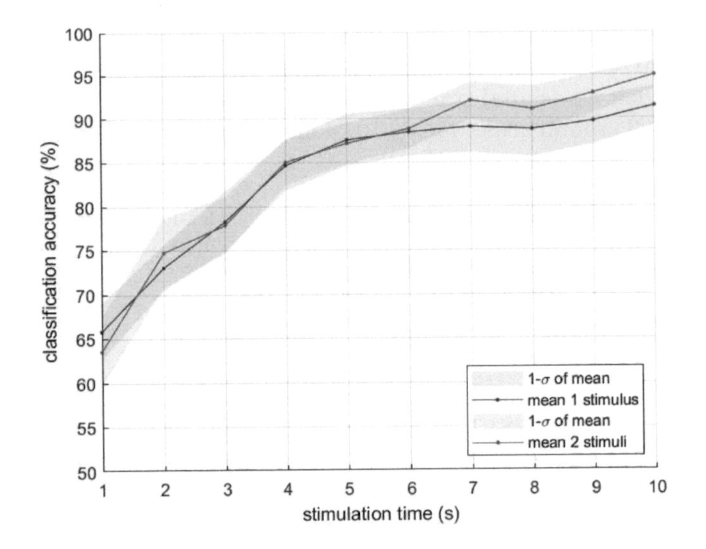

FIGURE 4.7 Mean classification versus SSVEP stimulation time (almost coincident with system latency). The standard deviation of the mean is also reported as a shaded area. The "1 stimulus" case is compared with the "2 stimuli" one.

25th percentiles. It can be seen that there is much dispersion around the median classification performance, but in the best cases the accuracy reaches 100 % in a few seconds. Also note that, while the mean and the median accuracies are almost identical at 1 s, the median is typically higher for longer stimulations. This indicates that the accuracy enhancement is actually higher for most subjects, while the performance remains poor for a few subjects. This was somehow anticipated by the results of Tabs. 4.1 and 4.2.

This reasoning was repeated in the case of four PSD features and the results are resumed in Fig. 4.9. These plots confirm that the classification performance is slightly better, but this enhancement is not statistically relevant. Nonetheless, more homogeneous performances can be obtained in some cases, notably increasing the accuracies for the worst subjects. As a proof of that, one can refer to the blue shaded area in Fig. 4.9(a), noticeably narrower than the one in Fig. 4.8.

Finally, it is useful to express the SSVEP-BCI performance in terms of the ITR from Eq. (4.7). Note that with $N = 2$ possible choices, the maximum theoretical ITR equals 60 bit/min if at least a 1 s-long stimulation is considered. Then, indeed higher ITRs could be reached for smaller system latencies, but this seems quite unfeasible in the current setting. The ITRs distribution among the subjects is shown in Fig. 4.10 for the case "1 stimulus" as a function of stimulation time. The box-plots show that only rarely some subjects overcome 20 bit/min thanks to high accuracies at short stimulation times. In

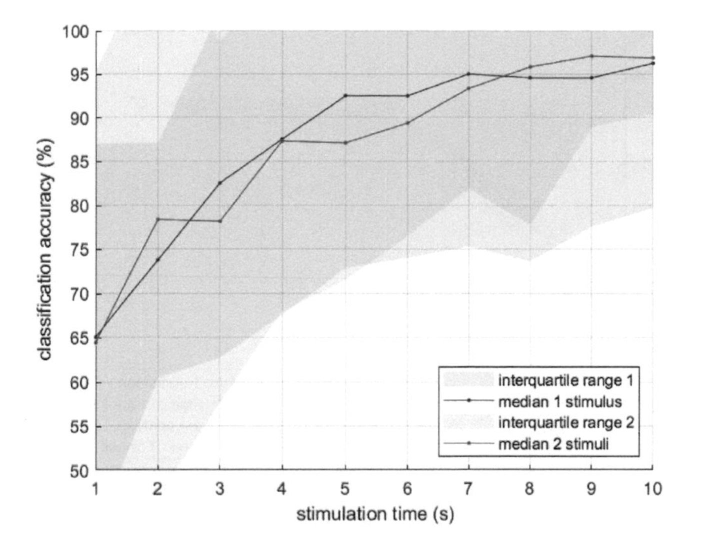

FIGURE 4.8 Median classification versus SSVEP stimulation time (almost coincident with system latency). The interquartile range is also reported with a shaded area as a measure of dispersion around the median. The cases "1 stimulus" and "2 stimuli" is compared.

this example, the optimal median ITR (red lines in the boxes) is at about 4 s to 5 s and it equals 7.8 bit/min, though some subject can perform better. Then, the ITRs are more homogeneous (i.e., there is less dispersion) for higher stimulation times, although the median value decreases.

4.3.2 Time Domain

In this subsection, the performance of the SSVEP time algorithm is assessed in terms of accuracy and response time.

Experimental setup

Twenty healthy subjects are involved in the algorithm validation phase. At first, the EEG signals of the subjects are acquired during free eye blinking. Then, subjects are asked to stare for 10 s, one at a time, at two stimuli on a screen that flickers at a frequency of 10 Hz and 12 Hz, respectively. These frequencies are sub-multiples of the refresh rate of the head-mounted display Epson Moverio BT 200 [138]. Moreover, according to previous studies, no training is required at these frequencies [110]. The optimal parameters for eye blinking detection and frequencies are found and, finally, the subjects is asked to perform a series of tasks with a robot. The task consists of turning the robot to the left or to the right, making it go straight on, or stopping. The actual operation will be detailed later.

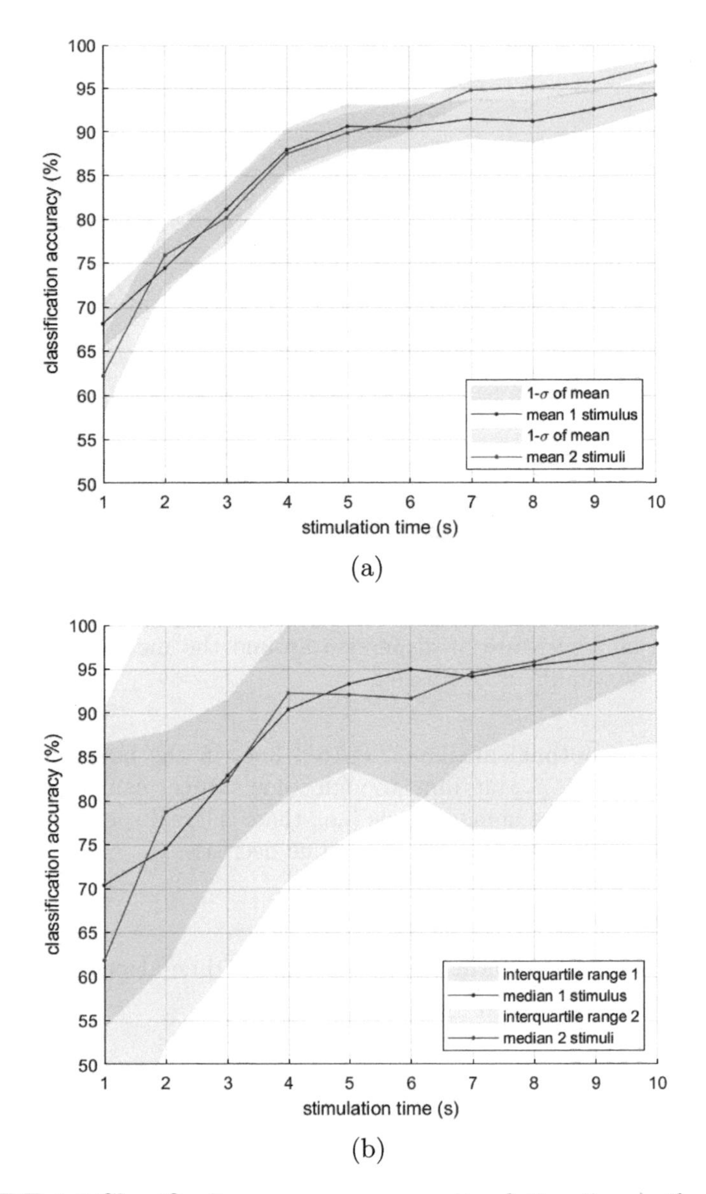

FIGURE 4.9 Classification accuracy versus stimulation time in the four PSD features case. Mean and standard deviation (a) and median and interquartile range (b).

Accuracy

The performance of the system is assessed both in relation to its ability to recognize the frequency of interest and in relation to the algorithm's ability

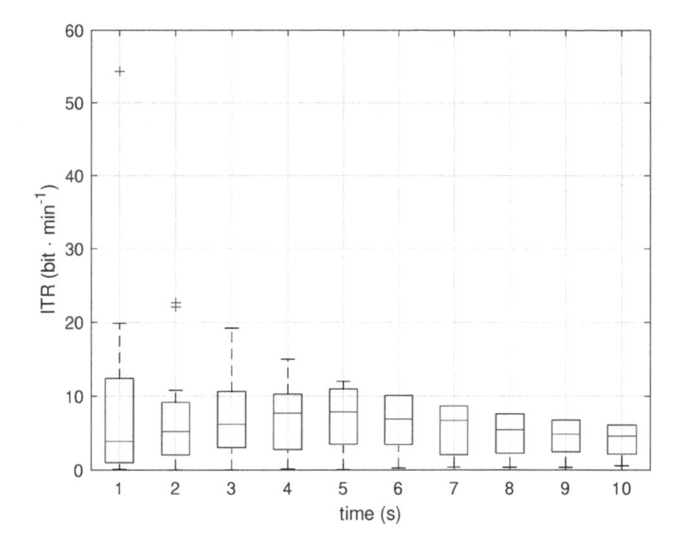

FIGURE 4.10 Box-plots of the Information Transfer Rate (ITR) as a function of SSVEP stimulation time (almost coincident with system latency). The case "1 stimulus" is here considered. The median values are reported as red lines in the boxes. The few outliers are associated with subjects that have quite higher performance in comparison with the median one.

to detect unintentional eye blinking. Finally, evaluation of the accuracy of the system as a whole is carried out.

To assess the accuracy of the algorithm, i.e., the ratio between the percentage of correct responses and the total ones, the features $F1$ and $F2$ are computed for different values of T: 0.4, 0.5, 0.8, and 1.0 s and different thresholds (varying in range 0.4 to 0.5). Results, in terms of accuracy and latency, are shown in Tabs. 4.3, and 4.4, respectively. Increasing *T1*, the accuracy also increases but the response time becomes longer. A trade-off must therefore be found between accuracy and time response (Fig. 4.11) The trade-off was identified at a response time of 1.5 s.

TABLE 4.3
Accuracy (%) of SSVEP detection algorithm for different time windows T and threshold values T1 [140].

		T1					
		0.40	**0.42**	**0.44**	**0.46**	**0.48**	**0.50**
T (s)	**0.4**	70.8 ± 6.7	70.8 ± 6.7	70.2 ± 7.3	69.3 ± 7.5	69.1 ± 7.6	69.7 ± 8.1
	0.5	73.1 ± 8.2	73.3 ± 7.8	74.7 ± 8.0	75.4 ± 7.6	77.9 ± 7.1	78.5 ± 6.4
	0.8	82.0 ± 5.8	84.0 ± 5.8	84.9 ± 5.4	86.1 ± 5.0	86.9 ± 5.2	88.1 ± 4.8
	1.0	86.0 ± 4.2	87.2 ± 4.1	89.4 ± 4.1	91.0 ± 4.2	91.8 ± 3.7	92.6 ± 3.6

TABLE 4.4
Time response (s) of SSVEP detection algorithm for different time windows T and threshold values T1 [140].

		T1					
		0.40	**0.42**	**0.44**	**0.46**	**0.48**	**0.50**
T (s)	**0.4**	0.64 ± 0.14	0.67 ± 0.15	0.72 ± 0.18	0.77 ± 0.19	0.81 ± 0.22	0.89 ± 0.26
	0.5	0.84 ± 0.18	0.89 ± 0.20	0.95 ± 0.22	1.04 ± 0.26	1.13 ± 0.31	1.22 ± 0.33
	0.8	1.65 ± 0.42	1.78 ± 0.45	1.94 ± 0.52	2.12 ± 0.56	2.37 ± 0.65	2.63 ± 0.73
	1.0	2.21 ± 0.54	2.41 ± 0.60	2.68 ± 0.69	2.99 ± 0.76	3.27 ± 0.84	3.71 ± 0.92

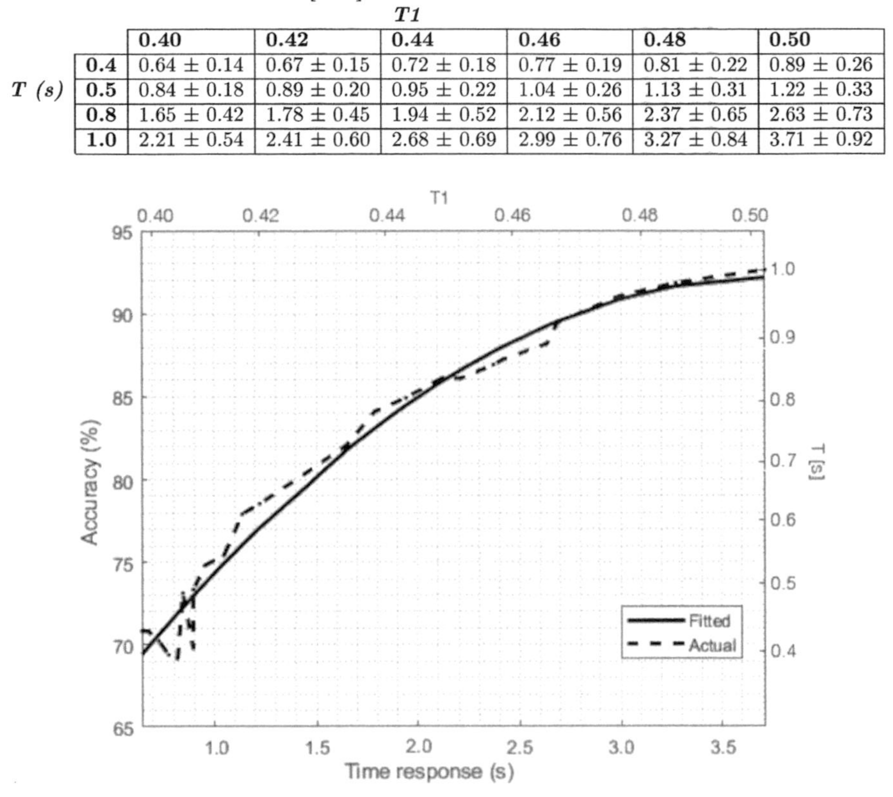

FIGURE 4.11 Trade-off between accuracy and time response highlighted in a plot vs parameters T1 and T [140].

The accuracy of voluntary eye-blink detection algorithm is defined the difference between the total number of eye blinking N_{blink} and those incorrectly classified E_{blink} and N_{blink} expressed as a percentage as shown in Eq. 4.8.

$$A_{subject} = \frac{N_{blink} - E_{blink}}{N_{blink}} \cdot 100 \qquad (\%) \tag{4.8}$$

The algorithm is validated on 10 subjects. Each participant is asked to blink voluntarily 10 times. Their EEG signal is acquired for 120 s so as to ensure that unintentional eye blinking occurs [146]. An accuracy of (91.8 ± 3.7) % is achieved.

Finally, the performance of the SSVEP/Eye blink integrated system is evaluated by using a Java application. This application simulates the control of the robot. The integrated algorithm is validated on 10 subjects. Each subject is asked to guide the robot in a labyrinth (Fig. 4.12). The *T* and *T1* are fixed to 0.50 and 0.44 s, respectively. Accuracy is evaluated as the ratio between the total commands sent to the robot $N_{commands}$ minus the number of incorrect commands $N_{commands}$ and $N_{commands}$ expressed as a percentage (Eq. 4.9)

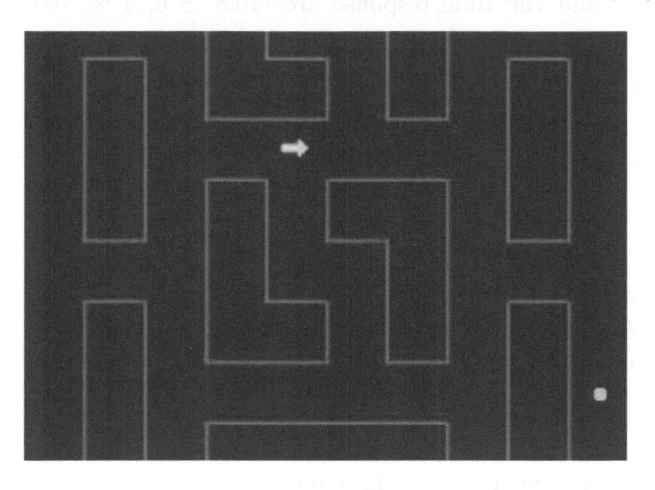

FIGURE 4.12 Java Application for simulating the behavior of the robot [140].

$$A_{subject} = \frac{N_{commands} - E_{commands}}{N_{commands}} \cdot 100 \qquad (\%) \tag{4.9}$$

The results are shown in Tab. 4.5.

TABLE 4.5
Performance of SSVEP/Eye blink integrated detection algorithm for T=0.5 s, $T1$=0.44, and eye blink threshold equal to 0 [140].

SUBJECT	Time (s)	Commands	Errors	Accuracy (%)
#1	124.91	30	5	83.3
#2	217.76	57	12	78.9
#3	107.44	33	6	81.8
#4	94.67	23	5	78.2
#5	102.98	33	5	84.8
#6	195.53	51	11	78.4
#7	59.23	16	2	87.5
#8	95.04	26	2	92.3
#9	104.34	32	4	87.5
#10	142.55	41	7	82.9
AVERAGE	124.44 ± 30.72	34 ± 8	6 ± 2	83.5 ± 2.9

4.3.3 Response Time

The other metric to assess system performance is the ITR [147]. Table 4.6 shows that ITR reached the maximum value for T= 0.4 s and $T1$= 0.40. The accuracy and the time response are (70.8 ± 6.7) %, (0.64 ± 0.14) s, respectively.

TABLE 4.6
ITR (bit/min) of SSVEP/Eye blink detection algorithm at varying time windows T and thresholds $T1$ [140].

		T1					
		0.40	**0.42**	**0.44**	**0.46**	**0.48**	**0.50**
T (s)	**0.4**	39.37	37.64	33.67	30.20	28.25	26.75
	0.5	33.86	32.25	32.66	30.65	31.80	30.46
	0.8	26.43	26.64	25.39	24.43	22.61	21.45
	1.0	23.43	22.46	22.16	21.23	20.06	18.27

4.3.4 Comparison with Literature

The SSVEP-based BCI discussed in this chapter is designed in order to be low-cost, wearable, and possibly trainingless in order to be closer to daily-life applications. The performance of the system was assessed by means of the experimental results shown above. However, to highlight the contribution of the presented solutions to the ongoing development in this field, it is essential to compare these results with the literature state of art. To this aim, some representative examples are here reported by mainly taking into account the last five years of research and development on non-invasive SSVEP-BCI. Moreover, another key point is the use of dry electrodes, which are receiving more interest in recent works but they pose some metrological challenges. In this regard, a useful summary is reported in [148].

Focusing on SSVEP literature, it is highlighted that detection algorithms are based on FFT, PSD, or Canonical Correlation Analysis (CCA). System calibration is not typically needed but it can be exploited to increase performance. As an example, the authors of [148] propose a variant of CCA that requires task-related calibration. The classification accuracies of those works are in the 83 % to 93 % range, while the ITR goes from 14 bit/min to 92 bit/min. It is worth noting that the highest ITR is obtained with the task-related CCA and by exploiting eight dry electrodes. Meanwhile, the ITR reported for the SSVEP-BCI with one electrode are about 38 bit/min [149, 150].

SSVEP has been largely exploited in building " mental spellers". In [107], a BCI employing three stimuli and single-channel EEG was proposed. As a performance indicator, a classification accuracy equal to 99.2 % was reported for a 10 s-long stimulation, and the calculated ITR was about 67 bitmin. However, the reproducibility of this result is not foreseeable since only five subjects participated in the experiments. Moreover, the system was not completely trainingless since there was the need to optimize the stimulation per each subject. A further speller based on single-channel EEG acquisition and dry electrodes was proposed in [118]. In this case, a Deep Neural Network was used and the classification accuracy reached 97.4 % with a 2 s-long visual stimulation. The assessed information transfer rate was 49.0 ± 7.7 bit/min, but only eight subjects were considered in the experiments and there is no indication about the cost of the setup. It is also clear that much data is required for the training of the Deep Net.

With specific regard to the combination of BCI with AR glasses, researches on VEP-based paradigms were recently surveyed [113]. By focusing on SSVEP-based works, it was reported that video see-through technology is mostly exploited instead of optical see-through. Moreover, the main applications are in the robotic field. A system integrating Smart Glasses and SSVEP-BCI is reported in [116]. The system was proposed for quadcopter control. However, in this system a total of 16 electrodes were employed, and the accuracy achieved while executing a flight task was 85 % (on only five subjects). The quite low accuracy points out that there are still many issues to face before the exploitation of such technology outside laboratories. Indeed, motion artifacts [115], which are probably caused by the flight task, affect the SSVEP detection. Furthermore, proper focusing on the flickering icon is important for eliciting SSVEP. In comparison with these literature works, the system illustrated here optimizes the wearability thanks to the use of a single-channel BCI, and it is relatively low-cost while reaching a classification performance compatible with literature ones. However, it is to note that even this SSVEP system was tested in a laboratory environment: though the setup is optimized for real-life applications, issues like motion artifacts were avoided by asking the subject to limit movements.

Many improvements are thus possible for this SSVEP-based BCI. The ITR should be increased while continuing to exploit the user-friendliness of a single-channel setup. Moreover, the EEG electrodes placement must be stabilized in order to use the system even during movements (portability). Especially for the present system, the EEG quality is essential for the system performance, while the processing algorithm can be based on a relatively simple approach. Finally, following an interesting suggestion from [103], the algorithm could exploit a dynamic stop to further optimize the latency/accuracy trade-off: such an algorithm would classify the EEG in real-time and stop when a certain degree of confidence is reached on the assigned class; in doing so, the time window would be variable, and possibly it would be the shorter possible per each task. Indeed, a critical part of such an approach would be the stopping criterion, which must limit the possible misclassifications. Some applications for the SSVEP system at the current development state are reported in the next chapter.

5

Case Studies

In this chapter, the applications for a SSVEP-based BCI tested with wearable system are described in order to show its possibilities in fields such as Industry 4.0 [151] and healthcare.

5.1 INDUSTRIAL MAINTENANCE

The possibility of successfully employing a wearable SSVEP-BCI is first investigated for a *hands-free* inspection task in an industrial context [31]. *Hands-free* operation is referred to the use of a device with no or limited manual manipulation. The scenario consists of a technician conducting inspection and maintenance of a complex industrial plant. In checking its parts, the technician can fix cables, electrical machines, or power drives while simultaneously visualizing useful information on Smart Glasses. Hands-free operation can thus provide an interesting feature: for instance, the user can scroll textual information without interrupting the task, or he/she can access sensor data if the industrial plant is equipped with transducers.

Figure 5.1 shows the SSVEP-BCI system communicating with a wireless sensor network for advanced diagnostics and real-time monitoring of an industrial plant. Bluetooth is exploited for communication between the Smart Glasses and the sensor network.

An Android application is purposely developed to scan available smart transducers and then connect to the one of interest. Therefore, the user is prompted to a transducer selection window. Once connected, available data are made accessible and the user can ask for specific measures, such as temperature or hydraulic pressure. The Android application diagram is depicted in Fig. 5.2. The commands are sent to the transducer by merely staring at the corresponding icon, without the use of hands.

Fig. 5.3(a) shows a user wearing the SSVEP-BCI prototype during an emulated inspection task. As an example, temperature and humidity of an industrial oven could be selected by the icons appearing on the Smart Glasses display: the user stares at the desired flickering icon and, after a few seconds, the corresponding command is sent. Figure 5.3(b) is a representation of the user's view through the Smart Glasses. Note that, while focusing on the icon to

DOI: 10.1201/9781003263876-5

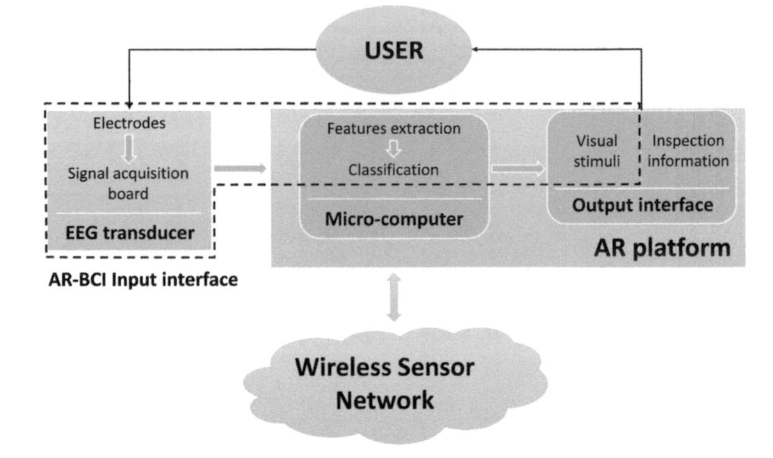

FIGURE 5.1 AR-BCI system based on SSVEP for accessing data from a wireless sensor network [31].

select, the background naturally appears blurred for the sake of the exposition clarity. Conversely, in actual conditions, they are correctly focused. In this way, the user can inspect the real environment while having superimposed textual information by means of the augmented reality glasses.

5.2 CIVIL ENGINEERING

A further noteworthy application for the wearable SSVEP-BCI is in inspecting bridges and viaducts. This application clearly belongs to the Industry 4.0 framework. Nevertheless, despite the similarities with the previous case study, it seems useful to discuss the peculiar aspects associated with practical applications in civil engineering.

In this field, optical-see-through devices appear useful for visual inspection, which is a crucial task when checking the integrity of a bridge or a viaduct. Hence, visualizing additional information about an infrastructure must not obstruct the visualization of the structure itself. The use of AR technologies in civil engineering is still at an early stage, but it is indeed appealing. In addition, the hands-free feature, provided in the present case by the BCI, guarantees easier navigation into the available information. Indeed the need to check the state of architectural works has always been an important and challenging issue, and the efforts in this direction have increased in recent years. A similar application can be found in the fruition of a cultural asset enriched by VR. Also in this case, the enjoyment of content projected on the

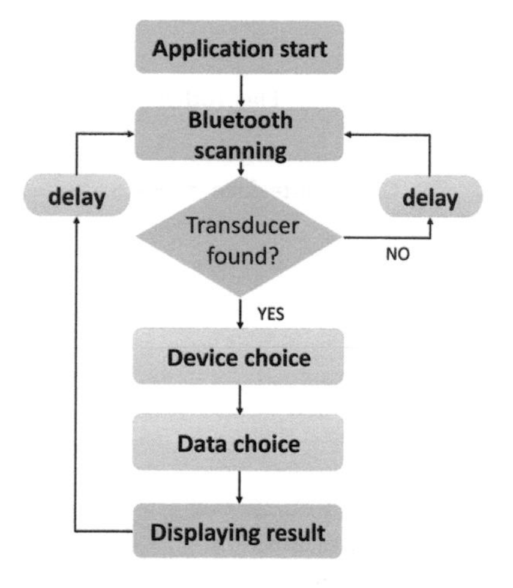

FIGURE 5.2 Android application diagram for an inspection task in industrial framework: smart transducers and related measures are selected by staring at corresponding flickering icons [31].

screen of the viewer can be facilitated by the use of BCI. With the wearable SSVEP-BCI, the attempt was to aid in this task.

For the civil engineering case study, the information consists of static and dynamic measures. In detail, accelerometers, gyroscopes, magnetometers, and strain gauges, as well as temperature and humidity sensors were installed on a beam to emulate measurements on a bridge within a laboratory setup. The technician is interested in monitoring these quantities during a load test in which deformations and vibrations must be measured.

A sketch of the SSVEP-based system communicating with a sensor stack is instead reported in Fig. 5.4. This clearly represents a novel possibility of interaction between the *Internet of Things* (IoT) and advanced monitoring systems. From a general point of view, the system follows the same implementation of the one previous described. Nonetheless, the need to continuously acquire and process the EEG arose since, differently from the previous case study, the application requires the possibility to asynchronously choose an icon/the measures to visualize. In doing that, the implementation of a variable acquisition window is attempted by considering the confidence associated with a class. Remarkably, this possibility is already mentioned at the end of subsection 4.3.4 as a future development. Unfortunately, a proper study on the acquisition stopping criterion is still missing and a rigorous analysis of such an implementation is addressed to future developments. Nonetheless, despite being in a very early stage of development, this principle is exploited in developing an Android application through which the BCI user can access the measures of interest.

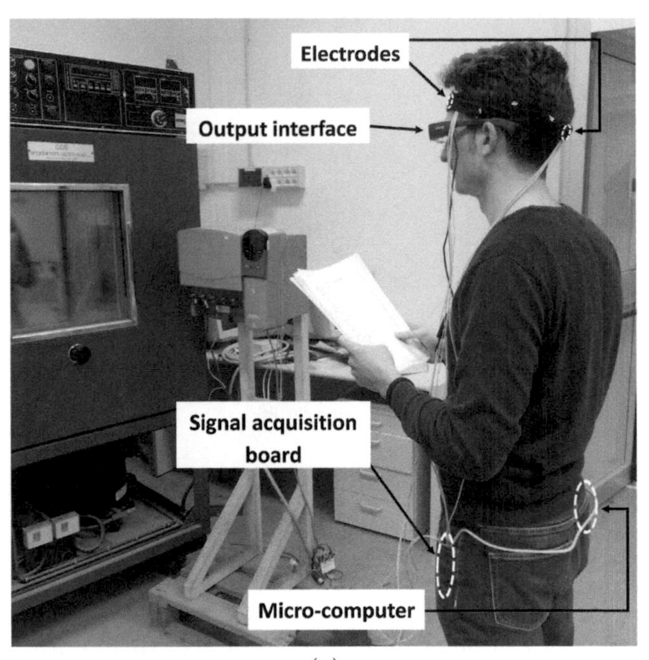

(a)

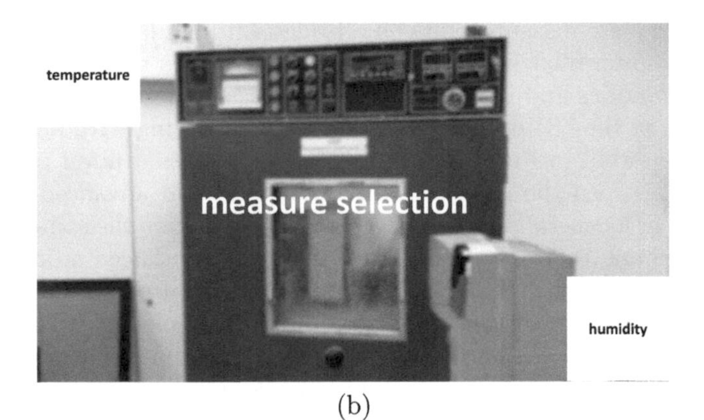

(b)

FIGURE 5.3 Example of inspection with the wearable SSVEP-BCI: measure selection menu, with flickering white squares for options (as usual in AR glasses, background image is blurred to focus on the selection). A user wearing the SSVEP-BCI system during an emulated inspection task (a) and Simulated measure selection window with two possible choices (b) [31].

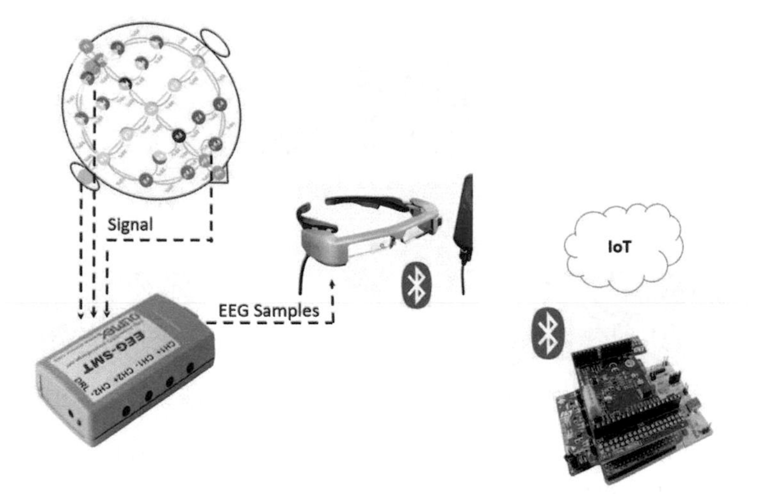

FIGURE 5.4 AR-BCI system based on SSVEP communicating with a sensor stack through Bluetooth low-energy.

The Android application for inspection tasks is customized to the civil engineering case study. The different activities are synthetically reported in Fig. 5.5. In there, the Smart Glasses simultaneously connect to the installed transducers, and then the user visualizes environmental measures and has the possibility to choose other data to visualize, either static measured or dynamic ones. The choice is conducted by means of flickering icons. In the current example, the static data consist of a deformation curve derived by merging strain gauges' measures and angle measures. Meanwhile, the dynamic data consist of vibration measures derived with 3D accelerometers.

5.3 ROBOTIC REHABILITATION

The SSVEP-BCI has also found applications in healthcare with particular regards to the rehabilitation of children with attention-deficit/hyperactivity disorder (ADHD) [152]. ADHD is a neurodevelopmental disorder in children and adolescents, characterized by hyperactivity, impulsivity, and an inability to concentrate. There are currently several treatments for ADHD, either cognitive, behavioral, or pharmacological [153]. Robot-assisted therapies are also currently available and the literature has shown that many individuals with cognitive deficits such autism or ADHD, especially children, prefer robotic-based therapies. In some circumstances they even respond faster when cued by robotic rather than human movement [154].

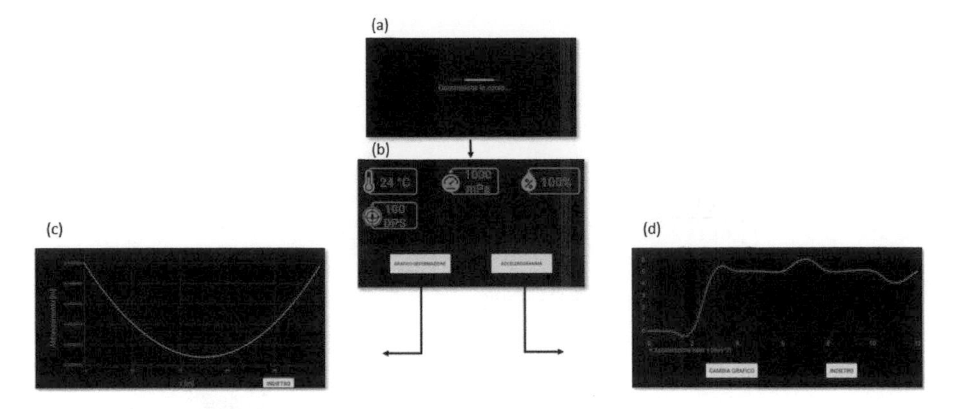

FIGURE 5.5 User interface in Android for bridge inspection during a load test. In the first activity, (a) the Smart Glasses automatically connect to multiple wireless transducers. In the second activity (b) the user visualizes environmental data. The third and fourth activities (c and d) can be accessed by selecting the data to visualize. The black background is equivalent to transparency when the application is running on Smart Glasses.

Indeed, the subjects feel more engaged by robot-assisted therapies, thus resulting in a greater commitment and increased levels of attention in the rehabilitation process, with a consequent improvement in its effectiveness [155]. In the field of neuro-rehabilitation, the combined use of eXtended Reality (XR) and BCI has been also used and it is becoming increasingly popular, precisely for children with ADHD. Therefore, the use of an integrated system including BCI, XR, and a robot for the treatment of ADHD is innovative but very promising [156], [157], [158], [159], [160].

The system guarantees the interaction with a robot through the wearable BCI to implement a behavioral therapy protocol. In this context, the Smart Glasses were still used to display the possible commands for the robot, while in future it is foreseen to also mix virtual objects with reality so as to achieve different levels of complexity in the interaction (mixed reality). At the current development state, the usefulness of the SSVEP-BCI relies on the fact that the children must focus on a flickering icon in order to interact with the robot. Therefore, the task cannot be conducted without paying attention, and this should enhance children's engagement in the therapy. Figure 5.6(a) depicts the robot to control during the interaction and the possible choices appearing on the Smart Glasses display [140].

An interesting feature is added to the system, i.e., the possibility to have a third control command through the detection of voluntary eye-blinks. Therefore, such a system can be defined as a hybrid BCI, since it integrates SSVEP with another type of control paradigm. Note that the eye-blink is typically an artifact from the point of view of EEG measurement, and involuntary

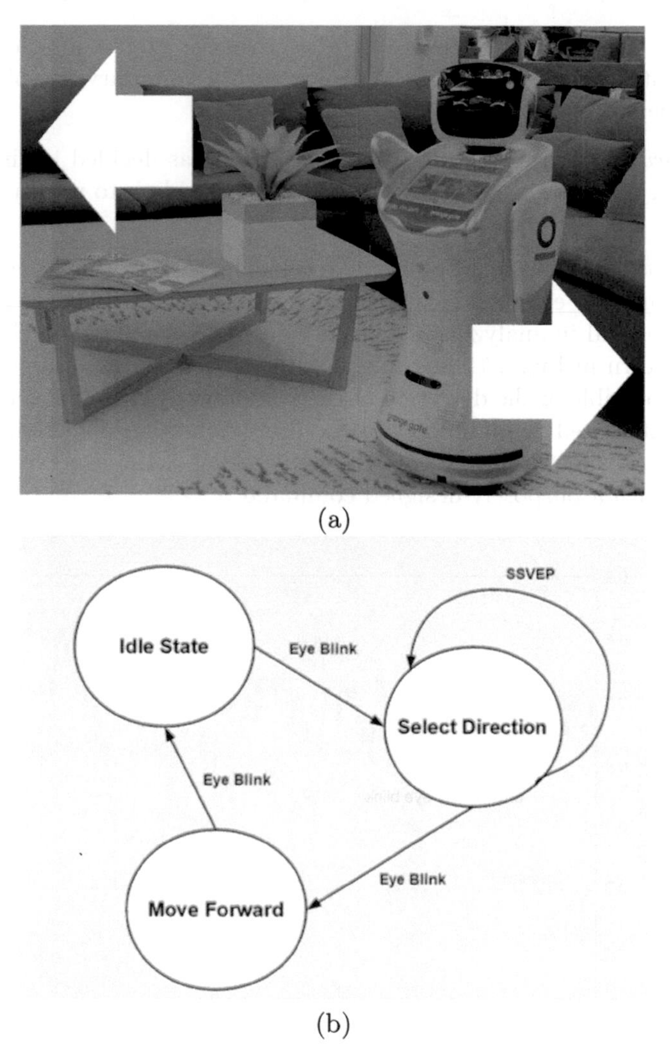

(a)

(b)

FIGURE 5.6 Application of the wearable SSVEP-BCI in the rehabilitation of children with ADHD through a behavioral therapy, as documented in [140]. Possible control commands in interacting with the robot (a) and Finite state machine implemented for robot control (b) [140].

eye-blinks were often present in measuring the brain signals with low-cost acquisition system. Nonetheless, in this case the artifact is exploited by implementing a finite state machine as follows (see Fig. 5.6(b):

- *idle state*: at the beginning of the interaction, no flickering icon is appearing and the robot is not moving; if a voluntary eye-blink is detected, the state is changed to select the direction;

- *selection direction*: through the SSVEP the user can choose left or right movement, and then confirm the choice with a voluntary eye-blink, which starts the movement;
- *move forward*: in this state the robot is moving as decided in the previous state and it can be stopped with a voluntary eye-blink to return to the idle state.

The distinction between voluntary and involuntary eye-blinks can be made by analyzing the eye-related artifact. This is actually done by choosing a proper threshold in analyzing the peaks related to eye-blinks in the time domain as shown in Fig. 5.7. The threshold is empirically determined. Clearly, errors are possible in the detection of the voluntary eye-blink and this further feature contributed to inaccuracy of the user's intention detection. However, such a feature allows to avoid continuous flickering of the icons, which can be activated with a purposely-designed command.

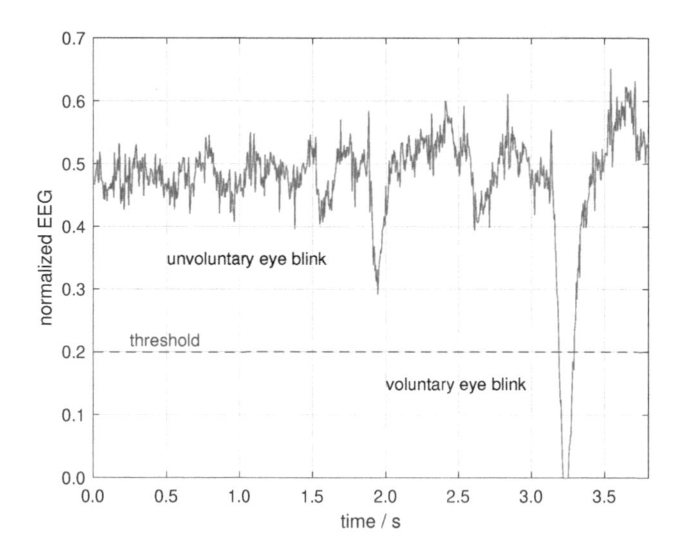

FIGURE 5.7 Distinction between voluntary and involuntary eye-blinks in time domain with an empirically determined threshold.

This system is validated on four children, aged between 6 and 8 years, with neuro-psychomotor disorders. Each child is asked to pilot the robot for 10 min. After reception of the command, the robot sends auditory or visual feedback signals to the subject. Experiments results are summarized in in Tab. 5.1.

The presented solution enables one of the main challenges of rehabilitation, namely ensuring the involvement and acceptability of the therapeutic setups, to be overcome. Children seemed to be very involved in the proposed exercises and the experiment showed that even an untrained subject could pilot the robot.

TABLE 5.1 Clinical case study: Performance of SSVEP/Eye blink integrated detection algorithm for T=0.5 s, $T1$=0.44, and eye blink threshold value equal to 0.

Subject	Age (years)	Presence of Parents	Initial Reluctance
#1	6	yes	no
#2	7	yes	yes
#3	6	yes	yes
#4	8	no	no

Part III

Passive Brain-Computer Interfaces

6

Fundamentals

This chapter introduces basic and complex mental states assessable by means of BCIs and relevant in the Industry 4.0 framework. In industry or in healthcare, humans and smart machines, are more and more connected through sensors of all kinds. Humans do not just exercise a defined role in an organization, but become part of a highly-composite automated system. In this context, the definitions of attention in rehabilitation, emotional valence in human-machine interaction, work-related stress, and engagement in learning and rehabilitation are provided in Section 6.1. In Section 6.2, for each of the considered mental states, the state of art on EEG markers is presented. Finally, in Section 6.3, the uncertainty sources in mental state measurements, as well as strategies to improve the experimental reproducibility, are discussed in accordance with a metrological perspective.

6.1 MEASURANDS

Passive BCI is the paradigm adopted when the user does not directly and consciously control his/her brain rhythms and, therefore, when the measuring goal consists just of monitoring the ongoing mental state [26]. Notably, electrical brainwaves are here treated by involving the EEG. Monitoring emotions, attention, engagement, and stress are pressing issues in several contexts. Nevertheless, these basic and complex mental states involve different theoretical aspects at varying application domains. This chapter presents fundamental definitions of mental states strictly linked to the Industry 4.0 framework.The ongoing technological transformation introduces new opportunities specifically related to an unprecedented kind of human-machine interaction. For instance, new adaptive automated rehabilitation platforms require real-time attention monitoring to improve their effectiveness. But also in Smart Industry, robots could be more and more collaborative if capable of monitoring the humans' stress.

On these premises, the following subsections present (Fig. 6.1): *attention in rehabilitation* (Section 6.1.1), *emotional valence in human-machine interaction* (Section 6.1.2), *work-related stress* (Section 6.1.3), and *engagement in learning and rehabilitation* (Section 6.1.4).

DOI: 10.1201/9781003263876-6

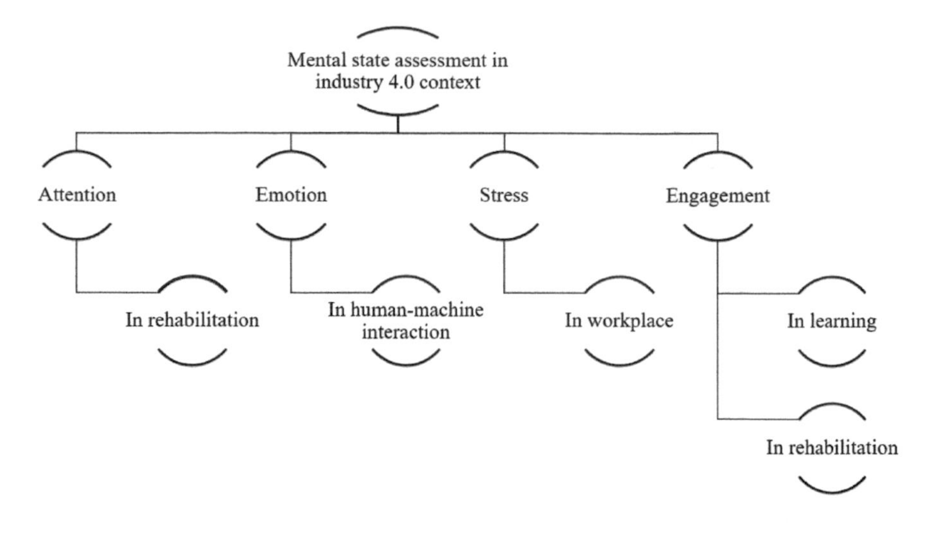

FIGURE 6.1 Mental state detection in the Industry 4.0.

6.1.1 Attention in Rehabilitation

In the framework of cognitive deficits, attention is defined as *the function that regulates the filtering and organization of the information received from a subject, allowing his/hers adequate responses* [161]. Indeed, attentional processes allow the subject to filter stimuli from the surrounding environment and to steer the cognitive resources to the stimulus of interest by ignoring other distracting ones.

As far as rehabilitation is concerned, four attentional dimensions can be considered [162]: (i) the *arousal* defines the psychophysiological activation required to allow perception of surrounding stimuli, (ii) the *selective attention* indicates the ability to focus on specific stimuli, (iii) the *distributed attention* is the capacity to process concurrent information at the same time, and (iv) the *sustained attention* is the ability to steer and maintain cognitive activity on specific stimuli.

The inability to focus attention on a stimulus and to ignore distractors underlies many cognitive deficits. Distractors thus play a fundamental role in studying the attentional processes [163]. Consequently, several studies on the measurement of attention are based on the use of distracting stimuli during the execution of cognitive or motor tasks.

In clinical practice, several tests are exploited for attention detection, e.g., *Go/No-Go tasks* [164], *Flanker tasks* [165], *Continuous Performance tests* [166], *Stroop test* [167], and *Attentional Performance tests* [168]. Biosignals are also used for attention detection: for instance, EEG and electrooculography are widely considered [169].

In the rehabilitation field, paying attention to the motor task increases effectiveness of rehabilitation processes [170]. Notably, Ang et al. proved how neuromotor rehabilitation exercises induce neuronal plasticity and promotes motor recovery [171]. Thus, repeating the exercise leads to a re-arrangement of the motor cortex. The repetition, however, can also produce tiredness in the subject and it may prevent an attentive focus on the execution of the exercise. Conversely, the execution of the exercise favors neuronal plasticity and motor learning when keeping the attentional focus in a sustained and selective way [172, 173].

6.1.2 Emotional Valence and Human-Machine Interaction

"Emotion" is derived from "emotus", a Latin word meaning *to bring out.* Emotion can be defined as *a complex set of interactions among subjective and objective factors, mediated by neural-hormonal systems, which can (i) give rise to affective experiences, such as feelings of arousal and pleasure/displeasure, (ii) generate cognitive processes, such as emotionally relevant perceptual effects, appraisals, and labeling processes, (iii) activate widespread physiological adjustments to the arousing conditions, and (iv) lead to behavior that is often, but not always, expressive, goal-directed, and adaptive* [174].

There are several theoretical models explaining emotions' nature. According to the *discrete theories*, a finite number of basic emotions exists. Ekman claims the existence of *primary emotions* like anger, disgust, fear, joy, sadness, and surprise [175]. Their intersection results in *secondary emotions. Dimensional theories* of emotion claim the existence of dimensions common to all emotions. As an example, the *Circumplex Model of Affect*, proposed by Russel and Mehrabian, argues that each emotion can be explained as a linear combination of three dimensions: valence (positive or negative), arousal (the already mentioned level of physiological activation), and dominance (i.e., presence/absence of control over the situation) [176].

Many theories in the emotions detection dwell on the activation of the brain during emotional expressions and, in particular, on the role of cortical brain lateralization. According to the *Theory of Right Hemisphere*, emotional expression and perception occur in the right hemisphere [177]. The right hemisphere also plays a relevant role for processing negative emotions according to the *Theory of Valence*, while positive emotions are mainly processed by the the left hemisphere [178]. The *Approach-Withdrawal Model* states that the left- and right-anterior regions have a role in processing emotional condition related to the control of approach and withdrawal behaviors [179]. Finally, the *Behavioral Activation System/Behavioral Inhibition System (BAS/BIS)* model claims that the left and the right regions are linked to the activation of the BAS and BIS systems, respectively [180].

The emotional system is involved in every human action, including computer-related activity. Monitoring the human emotional state during interaction with machines allows to optimize the interactions itself by adapting the machine behavior. Currently, in the scientific literature, real-time monitoring of emotions is proposed in application fields such as industry [181], health [182, 183], and entertainment [184].

6.1.3 Work-Related Stress

Stress is *a psycho-physical pathological condition in response to emotional, cognitive, or social tasks, which is perceived as excessive by a subject [185]*. When the homeostasis (i.e., self-regulating process by means of biological systems maintain stability of the organism by counteracting external or internal variation) and psychological well-being of an individual are altered by external stimuli (e.g., physical, emotional, etc.), a dysfunctional response can occur.

The identification of stress indicators is the first step in modeling and detecting stress. Psychological approaches, i.e., self-assessment scale, or questionnaires [186], and physiological parameters, such as cortisol concentration in blood or saliva, galvanic skin response, heart rate, and brain activity, are proposed in literature for stress assessment [187].

The first index of the individual's response to stress is the cortisol concentration. Cortisol is a hormone produced by the adrenal glands in order to save homeostasis in all the conditions altering the normal body balance. This is measured by means of invasive methods, such as repeated blood samples, or non-invasive but less reliable methods, such as saliva samples [188]. An additional indicator is skin conductance, a parameter associated to the activation of the sympathetic nervous system and correlated to stress [189]. In a stressful condition, the moisture in the epidermis of the individual increases and, therefore, skin resistance decreases. Moreover, stress produces peripheral vasoconstriction with a consequent increase in the heart rate [187].

Recently, brain activity was introduced as a further stress indicator. Brain activity is measured through fMRI, PET, or EEG. The latter method is the most widely used due to its easy implementation, mini-invasiveness, better latency, and robustness to artifacts due to physical activity [190, 191].

Stress detection is particularly significant in the industrial field: in the current fast-changing industrial world, technological innovation, has introduced new sources of stress, sometimes referred to as *Stress 4.0*. In the literature, several studies focus on the use of EEG-based systems for work-related stress detection, in order to improve workers' safety, health, well-being, and productivity [192, 190, 193, 194, 195].

Notably, in EEG-based systems, dry electrodes have been recently exploited because it was demonstrated that associated signal quality is comparable to wet sensing [196, 197], thus disclosing novel possibilities in industrial applications. In addition, the usage of dry electrodes has been shown to be particularly effective for reliably researching human cognitive states under real-world conditions. Nowadays, including an EEG device in workers' protective helmets during a construction process may allow brainwave monitoring and analysis, thus highlighting possible stress states and actual attention levels [198].

6.1.4 Engagement in Learning and Rehabilitation

The word *engagement* originated from the French verb "engager", meaning commit and/or involve. Numerous definitions of engagement have been proposed over time due to its multi-dimensional and heterogeneous nature. In 1990, Kahn defines this concept by referring to three dimensions: behavioral, cognitive, and emotional [199]. Behavioral engagement refers to observable signals (postures, gestures, actions, etc.) of persistence and participation. Cognitive engagement is the strain to extend intellectual commitment beyond the minimum necessary to complete a task. The third dimension, emotional involvement, is defined as the degree of positivity of an individual's emotional reaction to a task. Engagement is analyzed hereafter in the framework of learning activity and adaptive automated rehabilitation solutions.

Engagement and adaptive automated rehabilitation platforms

Engagement assessment is fundamental in clinical practice to personalize treatments and improve their effectiveness. Indeed, patients involved in healthcare decision-making tend to perform better and to be healthier. In rehabilitation, Graffigna et al. defined patient engagement as a *multi-dimensional psycho-social process, resulting from the conjoint cognitive, emotional, and behavioral enactment of individuals toward their health condition and management* [200]. The cognitive dimension refers to the meaning given by the patient to the disease, its treatments, its possible developments, and its monitoring. The emotional dimension consists of the emotive reactions of patients in adapting to the onset of the disease and the new living conditions connected to it. The behavioral dimension is connected to all the activity the patient acts out to face the disease and the treatments. Lequerica et al. defined engagement in rehabilitation as *a deliberate effort and commitment to working toward the goals of rehabilitation interventions, typically demonstrated through active, effortful participation in therapies and cooperation with treatment providers* [201].

Motivation plays an important role in feeding engagement. This can be extrinsic or intrinsic. Extrinsic motivation is linked to an external reward, such as money or praise, while in intrinsic motivation the rewards are internal to the person [202]. Intrinsic motivation is more powerful.

In rehabilitation, the main factors influencing patients' engagement are perception of the need for treatment, perception of the probability of a positive outcome, perception of self-efficacy in completing tasks, and re-evaluation of beliefs, attitudes and expectations [201].

In the case of pediatric rehabilitation, the situation is different. Indeed, it is not possible to rely on intrinsic motivation, but extrinsic motivation is required. Children react more easily to what is real, concrete, present, and immediately satisfying. Consequently, the relationship with the therapist becomes a fundamental extrinsic element in supporting the child's perceived self-efficacy in rehabilitation activity. This cognitive and emotional support is referred as *scaffolding* [203]. Thus, pediatric engagement is a concept "involving a connection, a sense of working together, and particular experiences that influence emotions, feelings, and motivation in the therapy process" [204].

Engagement in learning

In learning processes, automated learning platforms adapting in real time to the user's skills have been introduced with Industry 4.0. The philosopher Fred Newman defined learning engagement as *the student's psychological investment in and effort directed toward learning, under-standing, or mastering the knowledge, skills, or crafts that academic work is intended to promote* [205, 206]. Through the usage of the new immersive XR solutions and embedded *Artificial Intelligence* (*AI*), real-time human-machine interfaces allows a reciprocal adaptation between the subject and the machine [207, 208]. Nowadays, technological evolution is increasingly rapid and the users are continuously required to learn the use of new interfaces.

When the subject is learning, working memory identifies new information and long-term memory builds new mental schemes based on past ones [209, 210]. While pre-existing schemes decrease the load on working memory, the construction of new schemes increases it. Therefore, learning consists in building new schemes. Units of knowledge are related to different aspects of the world, including actions, objects, and abstract concepts [211]. When an individual learns a specific pattern, the neural structure of the brain is modified by the activation of the neuroplasticity process. During the learning process, new synaptic links are created. The brain builds a connection system of myelinated axons and the more the experience or task is repeated, the stronger the synaptic link between neurons becomes [212].

Therefore, the repetition of an activity is crucial in the learning process. When the subject learns new skills or modifies already learned ones, neuronal maps are created or enriched. Therefore, the introduction of new technologies requires a continuous readjustment of the brain's synaptic structures. The learning process is more effective the more the subject is engaged. An engaged user learns optimally, by avoiding distractions and increasing mental performance [213, 214].

6.2 STATE OF THE ART OF EEG MARKER IN PASSIVE BCI

In this section, EEG markers for mental state detection are presented along with a brief state of art of their implementation. The discussion considers *attention detection (Section 6.2.1), emotional valence assessment (Section*

6.2.2), stress monitoring (Section 6.2.3), and engagement recognition (Section 6.2.4).

6.2.1 Attention Detection

Many studies deal with assessing the attention and its different dimensions through the analysis of brain signals acquired with EEG [215]. Several studies have shown that the level of attention affects EEG signals [216, 217]. Therefore, variations in the EEG signal can be used to detect corresponding changes in attention levels [218]. Attention creates a variation in brain signals that can be assessed both in the time and frequency domains [219]. Different regions of the brain are activated during cognitive processes, and concurrent distracting events deactivate some brain areas while activating other ones [105].

In the field of rehabilitation, a personalized approach is commonly adopted in training a classification model in dealing with distraction detection. Some authors proposed a method to assess attention during ankle flexion-extension in the presence of auditory distractors [220]. Their method consists of extracting time-frequency features from the EEG recordings of 12 participants using an 18-channel system and wet electrodes. In doing that, they achieved an average accuracy of 71 %.

Another interesting study [221] proposed a processing system to detect attention during a cognitive task with eyes closed or open. An average accuracy of 69.2 % was achieved on five subjects with a 15-channel EEG system. Distraction detection during robot-assisted passive upper limb movements was proposed by Antelis et al. [222]. They achieved an average accuracy of 76.37 % in classifying 3-s epochs, when mentally counting backward in threes, from a self-selected three-digit random number, ensuring the distraction condition. The EEG signal was acquired from six patients with a 32-channel EEG using wet electrodes.

In 2019, Aliakbaryhosseinabadi et al. [220] presented an update of their previous work. An average accuracy of 85.8 % was obtained using a 28-channel EEG system and wet electrodes. The experimental setup involved three different distractors and exploits the motor-related cortical potential, and signal processing was based on spectro-temporal features extracted from 3-s epochs. However, the state of the art shows the lack of an appropriate approach for clinical applications. The high number of channels and the use of wet or semi-wet electrodes affect wearability, by limiting clinical usability.

6.2.2 Emotional Valence Assessment

Different biosignals have been analyzed for emotion assessment: *cerebral blood flow [223], electroculographic signals [224], electrocardiogram, blood volume pulse, galvanic skin response, respiration, and phalanx temperature [225].* Among these, the brain signal turned out to be promising. Brain signal can be measured through invasive and non-invasive techniques: PET, MEG, fNIRS, fMRI, and EEG. In particular, as mentioned above, EEG is characterized by an optimal temporal resolution which is particularly suitable for this kind of applications. In the prototyping of EEG solutions, a scientific challenge is to use dry electrodes [196, 197] and minimize the number of channels to optimize user comfort while preserving high performance.

In addition to dealing with the aforementioned reproducibility issues, measuring emotions requires the use of an *interval scale.* According to classification proposed by Stevens [226], four measurement scale can be defined: *nominal, ordinal, interval,* and *ratio.*

Nominal and ordinal scales represent non-additive quantities and, consequently, cannot be considered for measurements complying with the International Metrology Vocabulary [227]. The nominal scale is employed by studies adopting the *Theory of Discrete Emotions* [228], but it allows a classification of emotion. Conversely, measuring of emotions by means of interval scales, is made possible by adopting the *Circumplex Model.* Secondly, there is the reproducibility issue. The same stimulus or condition often induces different emotions in different individuals (loss of cross-subject reproducibility), but the same individual exposed to the same stimulus might react differently at different times (loss of within-subject reproducibility).

In psychology research, suitable sets of stimuli were validated experimentally by using significant samples and are widely used by clinicians and researchers [229]. In particular, several stimuli datasets were produced referring to the Circumplex Model and their scores were arranged along an interval scale. However, the problem of standardization of the induced response is still unresolved. In order to construct a reproducible metrological reference for measuring emotions, questionnaires or self-assessment scales can be used to test the effectiveness in inducing a certain emotion [230]. Furthermore, the use of such instruments during sample building can reduce potential emotional bias caused by psychiatric disorders.

EEG-based emotion studies have mainly concentrated on the asymmetrical behavior of the two cerebral hemispheres [231, 232, 233]. In particular, the relationship between emotions and brain asymmetry is theorized by two hypothesis models: the behavioral valence and the right hemisphere. The former hypothesis assumes that the right hemisphere is dominant over the left for all forms of emotional expression and perception. In contrast, according to the "valence theory", hemispheric asymmetry depends on emotional valence: the left hemisphere is dominant (in terms of signal amplitude) for positive emotions while the right hemisphere for negative ones. Specifically, the prefrontal cortex plays an important role in this theory because it controls cognitive functions and regulates the affective system [234].

Consequently, an index useful to evaluate the subject's emotional changes and responses and, therefore, to predict emotional states is the *frontal alpha asymmetry* (α_{asim}) [235]:

$$\alpha_{asim} = \ln(\alpha_{PSD_L}) - \ln(\alpha_{PSD_R}) \tag{6.1}$$

where α_{PSD_L} and α_{PSD_R} are the power spectral densities of the left and right hemispheres in the alpha band, respectively. This index could also predict emotion regulation difficulties by resting state EEG recordings. An advantage of this indicator is its robustness to individual differences [236].

The feature extraction procedure is carried out by Machine Learning algorithms. Therefore, also raw data from different domains (i.e., spatial, spectral or temporal) can be used as input to the classifier, namely, without an explicit hand-crafted feature extraction procedure. Spatial filters usually enhance sensitivity to particular brain sources, to improve source localization, and/or to suppress muscular or ocular artifacts [237]. Two different categories of spatial filters exist: those dependent on data and those not dependent on data. Spatial filters not dependent on data, such as Common Average Reference and Surface Laplacian spatial filters, generally use fixed geometric relationships to determine the weights of the transformation matrix. The data-dependent filters, although more complex, allow better results for specific applications because they are derived directly from user's data. They are particularly useful when little is known about specific brain activity or when there are conflicting theories (i.e., theory of valence and theory of the right hemisphere).

In the following, a state of the art of the main works related to emotion assessment is presented. This is also resumed in Tab. 6.1. Only works with at least an experimental sample of 10 subjects were included as having the minimum acceptable of statistically significant. The reported studies can be divided in two categories according to the used dataset: public and self-produced. Among the studies based on public datasets, those claiming the best accuracy on emotional valence detection use the datasets SEED [238, 239, 240, 241, 242, 243], DEAP [244, 245, 246, 247, 239, 240, 248, 249, 250, 251, 252, 241, 253, 254, 242], and DREAMER [251, 252, 243].

SJTU Emotion EEG Dataset (*SEED*) [255, 256] is a set of EEG data from the "Center for Brain-like Computing and Machine Intelligence" (BCMI lab) at Shanghai Jiao Tong University. Fifteen participants watched 15 videos of 4 s, arousing positive, neutral, or negative emotions while their EEG signals were recorded. No explanation needed for the selected videos, in order to employee an implicit emotion recognition task. EEG signals were acquired through the 62-channel Neuroscan system and, after each of 15 trials, participants compiled the self-assessment questionnaire to indicate their emotional feedback.

The Dataset for Emotion Analysis using EEG, physiological and video signals, mostly known as DEAP [257, 258], is composed by EEG signals acquired from 32 participants, during 40 videos lasting 1-min long music videos through the 32-channel system by BioSemi [259]. The emotion recognition task was implicit: no instruction was given during the experiment. Arousal, valence, like/dislike, dominance and familiarity were considered for each video.

DREAMER is defined by the authors as *The Database for Emotion Recognition through EEG and electrocardiogram Signals from Wireless Low-cost Off-the-Shelf Devices* [260, 261]. It is composed by EEG signals acquired during emotional elicitation through audio-visual stimuli using the 14-channel

Emotiv Epoc+ [63]. The emotional states, aroused during 18 film clips watching from 23 subjects, were amusement, excitement, happiness, calmness, anger, disgust, fear, sadness, and surprise. An implicit emotion recognition task was performed by participants who have been asked to rate their emotive states in terms of valence, arousal, and dominance. Song et al. [243] proposed a multi-channel EEG emotion recognition method based on a *Dynamical Graph Convolutional Neural Network* (*DGCCN*). Their study employed the 62-channels dataset SEED [262] and the 14-channels dataset DREAMER [260]. Within-subject and cross-subject settings on the SEED dataset achieves the average accuracies of 90.4 % and 79.95 %, respectively, in a three classes emotion recognition. On the DREAMER dataset, in the within-subject setting, an average accuracy of 86.23 % was achieved on valence dimension (positive or negative)

A multi-channel EEG-based emotion recognition [251] was realized by Liu et al. by employing a *Multi-Level Features guided Capsule Network* (*MLF-CapsNet*). They achieved an average accuracy of 97.97 % on the 32-channels DEAP on valence (positive or negative)classification, while an average accuracy of 94.59 % was reached on the 14-channels DREAMER dataset [257]. Similar accuracies were achieved by employing an end-to-end *Regional-Asymmetric Convolutional Neural Network* (*RACNN*) on the same datasets in a within-subject setup [252]. Studies based on self-produced datasets were presented [245, 263, 264] with a focus on studies using standardized image sets (i.e., *International Affective Picture System* (*IAPS*) [229], and *Geneva Affective Picture Database* (*GAPED*) [265]).

Mehmood et al., in their study, recorded EEG signals from 30 subjects with an 18 electrolyte gel-filled electrodes caps. They employed stimuli from the IAPS dataset to arouse positive or negative emotional valence [264]. The cross-subject accuracy of 70 % was achieved using the SVM classifier. Channel minimization strategies were also adopted from several studies in order to improve the wearability of the emotion detection device but, as a drawback, the use of a small number of channels implies a low spatial resolution. [266, 267, 268, 269, 270, 271, 272, 273, 274, 275]. Marín-Morales at al. focused on eliciting positive or negative valence trough virtual environments [274]. The EEG signal were acquired from 15 subjects by using a 10-channel device. Images from IAPS dataset were adopted as stimuli and each participant was subjected to the *Self-Assessment Manikin* (*SAM*) questionnaire to rate the effect of each stimulus. The SVM classifier achieved an accuracy of 71.21 % on EEG and Electrocardiogram (ECG) signals. Wei et al. used images from GAPED to elicit positive or negative valence in 12 participants while their EEG signal were recording by means of a headband coupled with printed dry electrodes [275]. A cross-subject accuracy of 64.73 % was achieved for two combinations of four channels. Within-subject accuracies of 91.75 % and 86.83 % were reached by the two combinations, respectively. Ang et al. elicted happy or sad emotions using pictures provided by the IAPS, by [272]. They involved 22 subjects. The EEG signal, acquired through *FP1* and *FP2* dry electrodes, was given in input to an *ANN* and an accuracy of 81.8 % was achieved. Each participant was administered self-assessment scales. In the study from Ogino et al., a single-channel EEG-based system to estimate valence [267] was developed. An accuracy of 72.40 % was reached, in the within-subject configuration, on EEG signals from 30 subjects. The processing step have provided for the application of Fast Fourier Transform, Robust Scaling and Support Vector Regression. SAM questionnaire was administered to the participants.

Pandey et al. proposed a single channel (*F4*) cross-subject emotion recognition system using Multilayer Perceptron Neural Network [271]. DEAP dataset was used an accuracy of 58.5 %. The study [273] presents an interesting data fusion technique for emotion classification using EEG, ECG, and *Photoplethysmogram* (*PPG*). The study refers to the discrete model of emotions which does not allow for a measurement but only for their classification. The EEG signals were recorded through an eight-channel device and were classified by a *Convolutional Neural Network* (*CNN*) achieving an average accuracy for the cross-subject case of 76.94 %. However, emotions were elicited by personal memories of the volunteers, non-standardized stimuli affecting the reproducibility of the experiment.

6.2.3 Stress Monitoring

Different tools are employed for stress detection: from psychological approaches like patient questionnaires and self-assessment scale [186], to physiological parameters like EEG signals, blood volume pulse, electrooculogram, salivary cortisol level [277], heart rate variability [278], galvanic skin response, or electromyography [187]. Among them, EEG signals exhibit reduced latency and less alteration due to physical activity (such as occurs for the heartbeat due to fatigue). [190, 191].

In the context of Industry 4.0, EEG has been widely employed to detect stress' workers [195, 194, 193]. Two approaches can be distinguished for EEG-based stress detection strategies: those based on *handcrafted features* and *data-driven* ones. The first approach aims to investigate the neurophysiological basis of the stress phenomenon, while the second one considers the brain as a black box. The two approaches are analyzed in the following subsections.

Methods based on handcrafted feature for stress detection

Systematic changes in frontal EEG asymmetry in response to emotional stimuli can be used to analyze emotional response. A positive relationship was found between frontal EEG asymmetry and concentrations of corticotropin-releasing hormone, which is involved in the response to stress, fear, anxiety, and depression. Thus, frontal EEG asymmetry can serve as an indicator for predicting emotional responses. For example, greater left relative frontal activity indicates greater self-reported happiness, whereas greater right relative frontal activity is associated with negative stimuli [279, 280]. However, the

TABLE 6.1 Studies on emotion recognition classified according to the employed datasets (i.e., SEED, DEAP, and DREAMER), stimuli (v="video", p="picture", m="memories"), task (i="implicit", e="explicit", n.a.="not available"), #channels, #participants, #classes, classifiers, and accuracies (n.a.="not available") [276].

Dataset	Study	Stimuli	Task	#channels	#participants	Classifier	#classes	Within-subject accuracy (%)	Cross-subject accuracy (%)
SEED	[238]	*v*	*i*	62	15	SincNet-R	3	94.5	90.0
	[243]	*v*	*i*	62	15	DGCNN	3	90.4	80.0
SEED & DEAP	[239]	*v*	*i*	62	15	DNN	3	n.a.	96.8
		v	*i*	32	32		2	n.a.	89.5
	[240]	*v*	*i*	62	15	SNN	3	n.a.	96.7
		v	*i*	32	32		2	n.a.	78.0
	[250]	*v*	*i*	62	15	SBSSVM	2	n.a.	72.0
		v	*i*	32	32		2	n.a.	89.0
	[241]	*v*	*i*	62	15	CNN	3	90.6	n.a.
		v	*i*	32	32		2	82.8	n.a.
	[242]	*v*	*i*	62	15	CNN	2	n.a.	86.6
		v	*i*	32	32		2	n.a.	72.8
DEAP	[244]	*v*	*i*	32	32	H-ATT-BGRU	2	n.a.	69.3
	[246]	*v*	*i*	32	32	CNN	2	n.a.	77.4
	[247]	*v*	*i*	4	32	LDA	2	n.a.	82.0
	[249]	*v*	*i*	32	32	LSTM-RNN	2	n.a.	81.1
	[253]	*v*	*i*	32	32	Kohonen-NN	2	76.3	n.a.
	[254]	*v*	*i*	32	32	SVM + FCM	2	78.4	n.a.
	[271]	*v*	*i*	1	32	MLP	2	n.a.	58.5
DEAP & DREAMER	[248]	*v*	*i*	32	32	BioCNN	2	83.1	n.a.
		v	*i*	14	23		2	56.0	n.a.
	[251]	*v*	*i*	32	32	MLF-CapsNet	2	98.0	n.a.
		v	*i*	14	23		2	94.6	n.a.
	[252]	*v*	*i*	32	32	RACNN	2	96,7	n.a.
		v	*i*	14	23		2	97,1	n.a.
SELF-PRODUCED	[243]	*v*	*i*	14	23	DGCNN	2	86.2	n.a.
	[245]	*v*	*i*	19	40	MLP, KNN, and SVM	2	n.a.	90.7
	[266]	*v*	*n.a.*	1	20	MC-LS-SVM	2	n.a.	90.6
	[263]	*v*	*n.a.*	14	10	RVM	2	91.2	n.a.
	[267]	*v*	*n.a.*	1	30	SVM	2	72.4	n.a.
	[269]	*v*	*i*	1	19	*k*-NN	3	94.1	n.a.
	[270]	*p*	*e*	3	16	SVM	6	n.a.	83.3
	[268]	*p*	*n.a.*	10	11	SVM	2	n.a.	84.2
	[264]	*p*	*n.a.*	18	30	SVM	2	n.a.	70.0
	[272]	*p*	*n.a.*	2	22	ANN	2	n.a.	81.8
	[274]	*p*	*n.a.*	10	38	SVM	2	n.a.	71.2
	[275]	*p*	*n.a.*	4	12	LDA	2	86.8	64.7
	[273]	*m*	*e*	8	20	CNN	3	n.a.	76.9

fear or happiness response to stimuli may be different in terms of frontal EEG asymmetry for different subjects [279].

According to Baron and Kenney's linear model, depending on the relative difference between the left and right hemispheres, EEG asymmetry may indicate either amplification or attenuation of the effect of fear stimuli. Some people show an increased relative activity of the left side compared to the right side in response to negative stimuli and an increased relative activity of the left side compared to the right side in response to positive stimuli. Frontal EEG asymmetry can also be a useful marker of depression and anxiety. Many works have demonstrated that stress induces changes in the prefrontal and frontal regions [277, 190, 279].

Data-driven methods for stress detection

In data-driven approaches, selected EEG features are those that maximize classification accuracy. EEG signals are subject to various problems, such as the low signal-to-noise ratio [23], non-stationarity in time domain [281, 282], and limited number of available data to identify a classification model [283, 284, 285].

To address these problems, different processing solutions and classification methods are implemented. EEG features extraction (in the time and frequency domains) and features selection are crucial steps for the accuracy and the computational cost of classification [8]. Temporal or frequency EEG features benefit from being extracted after spatial filtering, performed with either Principal Component Analysis (PCA) or ICA [286, 287]. The most used Supervised Learning algorithms to assess workers' stress using subjects' EEG signals are linear classifiers, such as LDA and SVM [288], ANNs, nonlinear Bayesian classifiers, kNN, RF, Naive Bayes, and Decision Tree [289, 290]. Matrix and tensor classifiers, and Deep Learning. The accuracy obtained with different classifiers, the number and type of electrodes used, and the different classes identified by different studies, are resumed in Tab. 6.2. The LDA is used to assess mental fatigue in [291], it divides the data into hyperplanes representing the different classes, with very-low computational burden. This is easy to use and provides satisfying results [292, 293, 294]. In [295], the SVM was used to identify three different levels of stress out of four though EEG features and six statistical features. The obtained recognition rate was 75.2 %. In addition, [296, 297] showed that combining different acquired signals allows for better accuracy (90.5 % and 92.0 %). Different Machine Learning algorithms were implemented in [195]: kNN, Gaussian Discriminant Analysis, SVM with different similarity functions (linear, Gaussian, cubic, and quadratic). Among the state-of-the-art classifiers, the SVM provided the highest classification accuracies: 90.1 % in [298] with a single-channel EEG and 88.0 % was reached by SVM in [299], where individuals' stress was assessed by exploiting only EEG signal as input of the classifier. State of art is summarized in Tab 6.2.

TABLE 6.2 State of art of stress classification [140].

Classifier	Reference	Reported Accuracy%	Acquired Signals	Classes	n° Electrodes	n° Subjects
Artificial Neural Network (ANN)	[296]	76.0 %	EEG,ECG,GSR	2 no-stress/stress	14 Wet	22
	[290]	79.2 %	EEG,SCL,BVP,PPG	2 levels of stress	5 Wet	15
Cellular Neural Network (CNN)	[296]	92.0 %	EEG,ECG,GSR	2 no-stress/stress	14 Wet	22
Decision Tree	[296]	84.0 %	EEG,ECG,GSR	2 no-stress/stress	14 Wet	22
Fisher linear Discriminant Analysis (FLDA)	[297]	90.5 %	EEG,EOG	2 alert and fatigue states	32 Wet	8
Gaussian Discriminant Analysis (GDA)	[195]	74.9 %	EEG,GSR	2 high or low stress level	14 Wet	11
K-Nearest Neighbors (**k-NN**)	[195]	65.8 %	EEG,GSR	2 high or low stress level	14 Wet	11
	[295]	76.7 %	EEG	2 levels of stress	14 Wet	9
Linear Discriminant Analysis (LDA)	[291]	77.5 %	EEG	3 low, medium, high mental fatigue	16 Wet	10
	[190]	86.0 %	EEG,ECG,EMG,GSR	3 stress,relax,and neutral	4 Wet	10
Naive Bayes (NB)	[296]	77.0 %	EEG,ECG,GSR	2 no-stress/stress	14 Wet	22
	[299]	69.7 %	EEG,ECG,GSR	2 mental workload and stress	2 Wet	9
Random Forest (RF)	[300]	79.6 %	EEG,EMG,ECG,GSR	4 cognitive states	8 Wet	12
	[289]	84.3 %	EEG,ECG,BVP	3 mental stress states	14 Wet	17
Support Vector Machine (SVM)	[295]	75.2 %	EEG	3 levels of stress	14 Wet	9
	[195]	80.3 %	EEG, SCL	2 high or low stress level	14 Wet	11
	[268]	85.4 %	EEG	2 positive or negative emotion	14 Wet	11
	[301]	87.5 %	EEG,ECG,HRV	2 stress and rest	2 Wet	7
	[298]	88.0 %	EEG	2 levels of stress	14 Wet	10
	[299]	90.1 %	EEG,ECG,GSR	2 mental workload and stress	2 Wet	9

6.2.4 Engagement Recognition

Two applications fields for engagement measurement are presented below. The first part of this subsection focuses on the *learning context*, while the second on the *pediatric neuro-motor rehabilitation* context.

Engagement assessment during learning activity

Different tools are proposed in literature for the detection of the three engagement dimensions (behavioral, emotional, and cognitive) in learning. Behavioral engagement is assessed through observation grids used to support direct observations or video analysis [302, 303]. Cognitive and emotional engagement are assessed through self-administrated questionnaires and surveys [304, 305]. In recent years, engagement assessment methods based on physiological biosignal have emerged: heart-rate variability, galvanic skin response, and EEG. Among the different biosignals, the EEG seems to be one of the most promising due to its low cost, low invasiveness, and high temporal resolution. A widely used EEG index in learning studies was proposed for the first time in 1995 by Pope. The case study was to decide when to switch from the autopilot to the manual mode during a fly simulation. This index is defined as:

$$E = \frac{\beta}{\theta + \alpha} \tag{6.2}$$

where α, β, and θ are the power spectral densities corresponding to the EEG frequency bands [8,13] Hz, [13,22] Hz, and [4,8] Hz, respectively. The main weakness of the abovementioned index is that it does not consider the three different engagement dimensions, but focuses only on the cognitive one. Afterward, EEG signals were proposed in several applications [306, 307] to include cognitive and emotional engagement assessment as well as the detection of underlying elements: emotions recognition and cognitive load activity assessment respectively [308, 309, 310, 311, 312, 313, 314].

One of the relevant metrological issues in the detection of engagement is the reproducibility of the experiments focused on emotional engagement. Indeed, the same stimulus does not necessarily induce the same emotion in different subjects. To build a metrological reference for reliable EEG-based emotional engagement detection, a possible solution is the combined use of standardized stimuli and subjects' self-assessments [230]. As concerns cognitive engagement, different cognitive states can be induced by increasing levels of difficulty of the proposed task and introducing complex schemes. When the subject is learning, working memory identifies new information and long-term memory builds new schemes based on past ones [209, 210]. While pre-existing schemes decrease the load on working memory, the construction of new schemes increases it. Therefore, learning consists in building new schemes: units of knowledge, each relating to different aspect of the world, including actions, objects, and abstract concepts [211]. When an individual learns a specific pattern, the neural structure of the brain is modified by the activation of the process neuroplasticity. During the learning process, new synaptic links are created: the brain builds a connection system of myelinated axons and, the more the experience or task is repeated, the stronger the synaptic link between neurons becomes [212]. Therefore, the repetition of activity is crucial in the learning process.

When the subject learns new skills and/or modifies already learned ones, neuronal maps are created and/or enriched. Thus, the cognitive engagement level grows up according to the proposed exercise difficulty increases.

Engagement detection in pediatric rehabilitation

Engagement in clinical practice is typically assessed trough questionnaires or rating scales. In adult rehabilitation, the most used are: *Patient Activation Measure (PAM-13) [315] and Patient Health Engagement (PHE) scale [316].* In pediatric rehabilitation, the *Pediatric Rehabilitation Intervention Measure of Engagement-Observation (PRIME-O)* version [317] and the *Pediatric Assessment of Rehabilitation Engagement (PARE)* scale [318] are used. Also in the rehabilitation context, the employment of bio-signals like eye-blinking [319], heart rate variability [320], and brain activity [321, 322] is an emerging trends for engagement detection. Most rehabilitation studies consider only cognitive dimension of engagement, neglecting emotional and behavioral ones. A EEG based-real-time cognitive engagement detection was proposed by [323]. In this study, a deep CNN was employed in a binary classification problem (engagement vs. disengagement detection). EEG signal was acquired from eight subjects during a *Go/No-Go task* and an average inter-subjective accuracy of 88.13 % was achieved. Cognitive engagement was also assessed in [324] during active and passive motor tasks' execution, an average classification accuracy of (80.7 ± 0.1) % for grasping movement and (82.8 ± 0.1) % for supination movement was achieved. The subjects involved in this study were all adults and, only recently, an EEG based engagement assessment study was proposed in pediatric rehabilitation. EEG signals were acquired from five children (some with cognitive disorder) and a inter-subjective accuracy of 95.8 % was reached in classifying positive or negative engagement without specifying the considered engagement dimensions [325].

6.3 STATEMENT OF THE METROLOGICAL PROBLEM

In general, several metrological problems need to be addressed for EEG-based measurement of mental states:

- *Measurands not uniquely defined*: several theories have been proposed to define the considered mental states. For instance, both the circumflex model and the discrete model have been introduced to define emotional states. Moreover, several definitions for engagement have been put forth by considering multiple dimensions. Attention is also a complex construct and several components or cognitive processes are considered for its definition.

- *Assessment tools not compatible with measurement objectives*: for example, in the case of the measurement of emotions, psychometricians use nominal scales that allow for emotion classification rather than their actual measurement.
- *The reproducibility question*: often, the same stimulus or condition induces different states in different individuals (loss of cross-subject reproducibility), but also the same individual exposed to the same stimulus at different times might react differently (loss of within-subject reproducibility). The usage of explicit influence tasks rather than implicit ones is another factor that affects the effectiveness of mood induction. Indeed, an explicit task promotes participants' entry into the target mental state but may be an additional cause of uncertainty. The existing standardized stimuli are mainly characterized in an implicit setup.
- *Lack of robust EEG-patterns for mental states*: different EEG features have been proposed for the identification of the same mental condition. Many of these characteristics have been considered statistically significant by some authors but not by others.
- *Lack of standardized stimuli for eliciting mental states*: apart from images, there is a lack of standardized stimuli to be used in experimental protocols. For example, currently, large standardized audiovisual databases are missing for emotion elicitation.

The following chapters present approaches to the mental state assessment based on metrologically rigorous methods. Theoretical models of the measurands were adopted compatible with the goal of measurement and not just with the goal of classification. Standardized stimuli were almost ever used in combination with self-assessment questionnaires. Experimental sample screening were realized before the campaigns in order mitigate the bias of psychological disorders.

7

EEG-Based Monitoring Instrumentation

This chapter presents the general characteristics of instrumentation for monitoring mental states. The considered instrumentation is based on EEG solutions implemented always in a highly wearable mode. In particular, the design of such an instrumentation is presented in Section 7.1, while the prototyping is discussed in Section 7.2. Both design and prototyping are detailed according to the specific mental state to be assessed.

7.1 DESIGN

As a first step, the system design is addressed by referring to both the basic ideas and the architecture.

7.1.1 Basic Ideas

The method for monitoring mental states is based on some key concepts:

- *High wearability*: the EEG acquisition system is ergonomic and comfortable. It is equipped with a rechargeable battery and transmits the acquired data via Bluetooth or Wi-Fi. Dry or semi-dry electrodes avoid the inconvenient of electrolytic gel.
- *Off-the-shelf components*: the measurement of mental states with low-density EEG (few channels) allows high wearability, maximum accuracy, and low latency by exploiting instrumentation available on the market [326].
- *Clinical applicability*: wearability cannot be a prejudice for accuracies compatible with clinical use.
- *An EEG-based method for mental state detection*: mental states are mediated by specific brain circuits and electrical waveforms. Therefore, the EEG signal varies according to the emotional, stress, attentional or engagement state of the subject. However, by using suitable algorithms, such a state can be recognized.

DOI: 10.1201/9781003263876-7

- *EEG-based subject-adaptative system*: new input channels (EEG) for intelligent interactive systems improve their user adaptability in the context of Industry 4.0.
- *Multi-factorial metrological reference*: systems are calibrated by using standardized strategies for inducing and assessing different mental states.

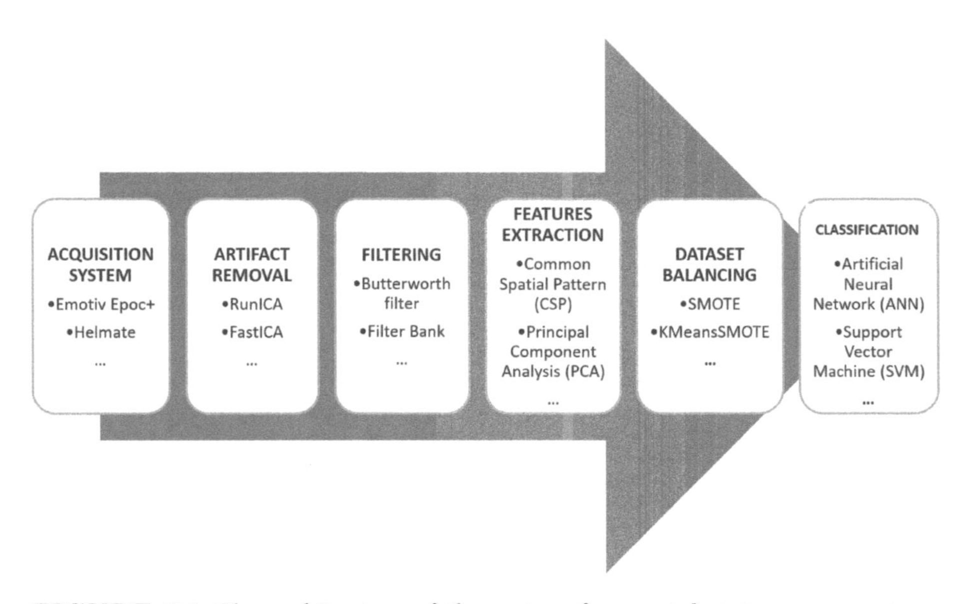

FIGURE 7.1 The architecture of the system for mental state assessment.

7.1.2 Architecture

The architecture of the system for mental state assessment is shown in Fig. 7.1. This is a general scheme to be customized to the specific application. The EEG signal is acquired by using *dry or semi-wet electrodes* from the scalp of the subject. According to the specific application, a different number of electrodes can be used and the material can vary from conductive rubber to silver. Namely, if the mental phenomenon of interest is originated in the prefrontal cortex, the only *FP1* and *FP2* electrodes are sufficient. Meanwhile, in other applications, a larger number of electrodes may be required. The signal is generally conditioned by stages of amplification and filtering (*analog filtering*), and then digitalized by the ADC. Moreover, it is eventually transmitted by the *Wireless Transmission Unit* for the *Data Analysis* stage.

Typically, artifacts removal algorithm are applied to the raw signals. The features are extracted from the signal with a feature extraction algorithm depending on the specific application. In case of unbalanced dataset, an oversampling procedure is also carried out to obtain the same amount of data in each class. New data are generated to be compatible with the statistical distribution of the experimental data already collected. The oversampling methods vary depending on the techniques adopted for the generation of the new data. The extracted features are finally given in input to a classifier for the mental state assessment. Different kinds of classifiers can be used, from standard Machine Learning techniques to advanced ANNs.

Hereafter, the general architecture is declined into specific applications by referring to the aforementioned application fields.

Attention in rehabilitation system architecture

The architecture specific for distraction detection in rehabilitation is shown in Fig. 7.2. Eight *active dry electrodes* acquired the EEG signals from the scalp. A 16.5-s trial followed by 16.5-s baseline period was acquired. The signals are conditioned by the *Analog Front End* and then digitized by the *Acquisition Unit* before transmission to the *Data Analysis* phase. In this stage, artifacts are removed by ICA and a *Filter Bank Common Spatial Pattern* (*FBCSP*) is implemented for the features extraction phase. Extracted features are finally given in input to a kNN classifier and attentional state is detected.

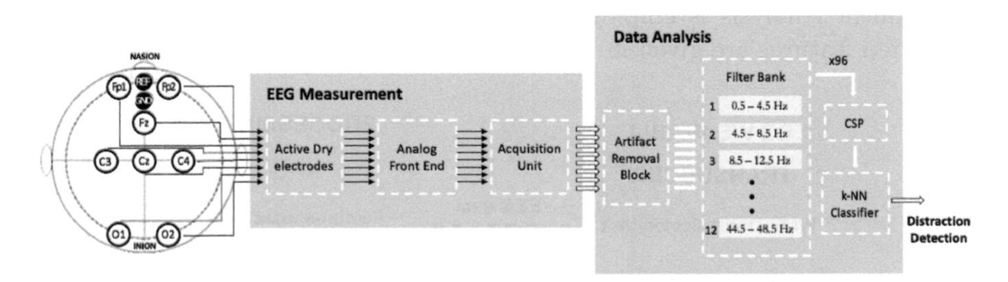

FIGURE 7.2 The system architecture specific for distraction-detection (CSP: Common Spatial Pattern algorithm)[307].

Emotional valence system architecture

The system architecture specific for emotional valence monitoring is shown in Fig. 7.3. The architecture is very similar to the previous one. Also in this case, the EEG signal is acquired by means of eight *active dry electrodes* with 30 s acquisition windows.

The main difference between the two architectures is the choice of classifier. In this solution, ANN is the adopted classifier.

Stress system architecture

Stress detection system architecture is shown in Fig. 7.4. The electrodes *FP1* and *FP2* acquired the EEG signals with reference to the earlobe. Acquisition windows of 180 s are set. Acquired signals are digitized by the *Acquisition*

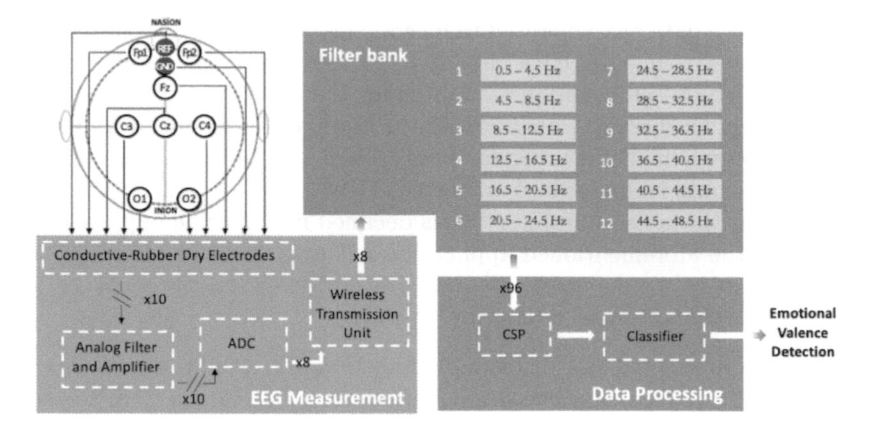

FIGURE 7.3 The system architecture specific for emotion-detection (CSP: Common Spatial Pattern algorithm) [307].

Unit and transmitted to the *Processing Unit* via wireless. Then, the Principal Component Analysis is employed for the feature extraction phase and the extracted features are given in input to the SVM for stress assessment.

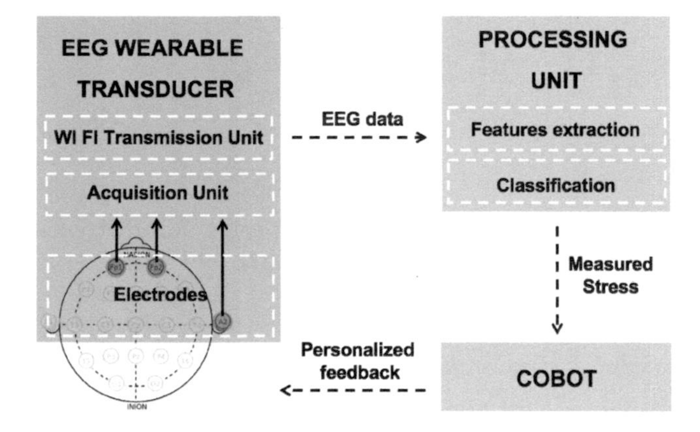

FIGURE 7.4 Architecture of the real-time stress monitoring system in Cobot interaction [140].

Engagement in rehabilitation

The architecture specific for engagement detection in rehabilitation is shown in Fig. 7.5. The EEG is acquired by 14 *semi-wet electrodes* referred to CMS/DRL with an acquisition time window of 15 min. The EEG measurement block is very similar to the previous cases while the *Data Processing* consists in (i) features extraction phase, (ii) dataset balancing, and (iii) classification phase in order to detect the cognitive and emotional engagement. Their intersection allows to define the engagement state.

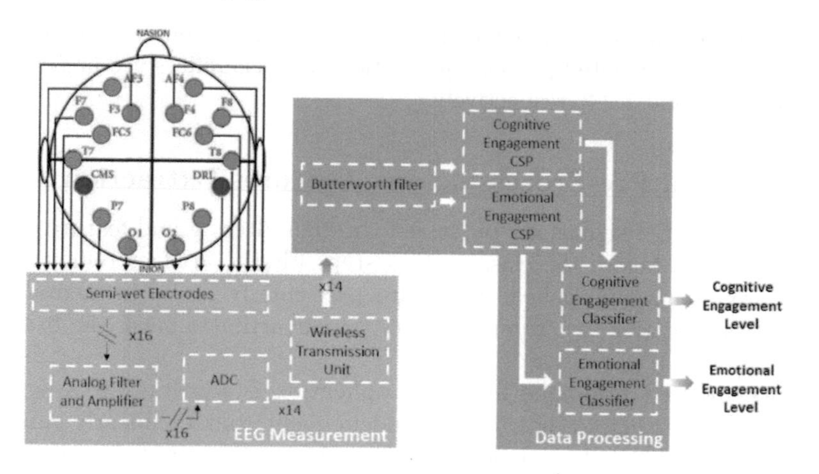

FIGURE 7.5 The architecture of the system for engagement assessment in rehabilitation [307].

Engagement in learning system architecture

The specific architecture of the system for engagement assessment in learning is depicted in Fig. 7.6. Signal acquisition and data processing phase are similar to the attention detection case but, after artifact removal, in the cross-subject case, a baseline removal and the *Transfer Component Analysis (TCA)* are employed in the training phase of the SVM classifier. Another difference is the duration of the acquisition window: 45 s of EEG acquisitions were collected for each participant.

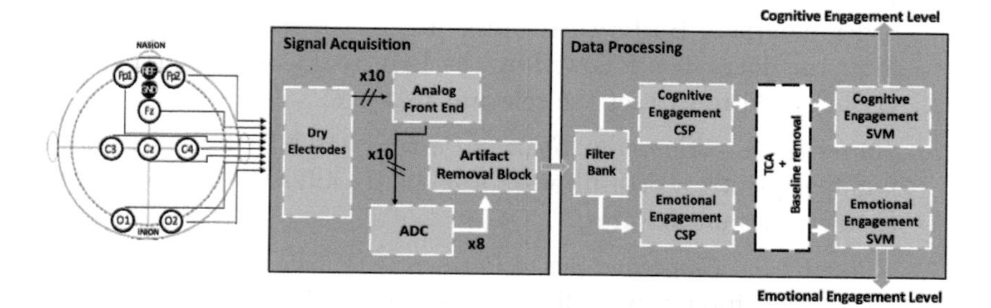

FIGURE 7.6 The architecture of the system for engagement assessment in learning; the white box is active only in the cross-subject case (ADC - Analog Digital Converter, CSP - Common Spatial Pattern, TCA - Transfer Component Analysis, and SVM - Support Vector Machine) [327].

7.2 PROTOTYPE

In accordance with the presented designs and with reference to these applications, the implemented processing solutions and classifiers are presented in Sections 7.2.1 and 7.2.2, respectively.

7.2.1 Signal Pre-Processing and Features Extraction

The acquired EEG signals are processed through filtering and feature extraction algorithms before the classification step. Firstly, artifact removal techniques are commonly applied. These are particularly useful in mental state detection application, where muscular or ocular artifacts mask the brain activity.

A widely-used technique in EEG application is the ICA. ICA a signal processing algorithm to discriminate independent sources linearly mixed in several sensors [328]. It can be applied to many different types of signals. In particular, ICA is one such method that deals with problems, which can be symbolized by the "cocktail-party-problem"; a number of microphones are situated at different positions in a room where a party is unfolding. In this room there are numerous sound sources: music being played, people talking, noises from the streets, etc. These sounds represent the *source signals.* The microphones acquire different mixtures of these sounds. Providing that there are at least as many microphones as original sources signals, the statistical independence and non-gaussianity of the source signals, ICA is able to extract the original sound sources from the set of mixtures, and the corresponding mixing proportions for each observed signal mixture

Generally speaking, four significant strategies can be mentioned for filtering and features extraction:

1. *Butterworth - Principal Component Analysis* (BPCA): a fourth-order bandpass Butterworth filter [0.5, 45.0] Hz was employed to filter the data, then followed by the Principal Component Analysis (PCA) [329] to extract the relevant features;
2. *Butterworth - Common Spatial Pattern* (BCSP): a fourth-order bandpass Butterworth filter [0.5, 45.0] Hz followed by the CSP to extract relevant features;
3. *Filter Bank - Common Spatial Pattern* (FBCSP): a 12 IIR bandpass Chebyshev filter type 2 filter bank with 4 Hz amplitude, equally spaced from 0.5 Hz to 48.5 Hz was applied to filter data, followed by a CSP algorithm to extract relevant features;
4. *Domain adaptation*: only in the cross-subject approach, a baseline removal and a TCA were adopted.

In details, the *PCA* works by approximating the signals as a linear combination of a few numbers of orthogonal components [330]. In doing so, data variance is outlined in an efficient way. PCA also performs a filter function, namely, maximum variance (information) of the data is highlighted. This eventually allows to improve signal-to-noise ratio through merely selecting the components with the greatest variability.

Meanwhile, the *CSP* algorithm is essentially used in binary discrimination between classes. It relies on the computation of the covariance matrices associated with the two considered classes, and it exploits their joint diagonalization to find an optimal spatial transformation for the EEG data. Thus, a transformation matrix is computed in order to project the data into a space where the differences between the class variances are maximized. The transformation matrix is also computed such that the projected components are sorted by variances in a descending order for the first class and ascending order for the second class [331].

To ease the classifier training phase, a domain adaptation can be also useful to address the problem of data variability between subjects. It should be recalled that this variability is due to the non-stationarity nature of the EEG signal. This problem makes the cross-subject approach a very challenging task [332]. Recently, domain adaptation methods [333] are of particular interest for the scientific community. In particular, the *Transfer Component Analysis* (TCA) [334] is an entrenched technique of domain adaptation already implemented in the EEG signal classification literature with promising results [332]. In case of two data domains, TCA works by computing a data projection ϕ that minimizes the *Maximum Mean Discrepancy* (MMD) between the data distributions, that is:

$$MMD = \left\| \frac{1}{n_S} \sum_{i=1}^{n_S} \phi(\vec{x}_{Si}) - \frac{1}{n_T} \sum_{i=1}^{n_T} \phi(\vec{x}_{Ti}) \right\|^2 \tag{7.1}$$

where n_S and n_T are the numbers of points in the *source* and *target* domains, respectively, while $\vec{x}_{S_i}$ and $\vec{x}_{Ti}$ are the $i-$th point (epoch) in the respective domains. The data are projected in a new latent space and then they are given as input for the classification stage. However, a domain composed of several sub-domains is usually generated by the different N subjects or sessions. In these cases, TCA can be applied by considering for the first domain a subset of samples from $N-1$ subjects, and with the data of the remaining subject for the other domain [332]. Unfortunately, this technique does not consider the fact that different subjects may belong to quite different domains, and it leads to poor results. A simple way to address this issue could be to subtract to each subject a baseline signal acquired from the subject himself/herself.

Finally, in case of unbalanced dataset, an oversampling method can be applied to obtain the same amount of data for each class. As above mentioned, oversampling methods allows the generation of new data compatible with the statistical distribution of the available data. The reason behind the use of an oversampling method is that, in the case of underrepresented classes, a classifier would implement a decision rule in favor of the major class during the training phase. This could happen because a classifier learns the most effective rule to maximize its performance, for instance its classification accuracy. Examples of *oversampling techniques* include: SMOTE [335], BorderlineSMOTE [336], ADASYN [337], SVMSMOTE [338], and KMeansSMOTE [339]. The various methods differ depending on the technique used to generate the new data. In particular, *Synthetic Minority Over-sampling Technique* (SMOTE) generates synthetic data by interpolating between observations in the original dataset: given a point in the dataset, a new observation is generated by interpolating between the point and one of its neighbors. ADASYN is similar to SMOTE but the number of samples generated for a given point depend on the number of nearby samples not belonging to the same class as the point. BorderlineSMOTE generates synthetic values in the border area between the classes. Similarly, SVMSMOTE creates new data by employing an SVM to find the optimal boundary between the classes. Instead, KMeansSMOTE works by selecting the best domain areas to oversample by using the k-means [340] algorithm.

7.2.2 Classification

The classification step usually involves a performance comparison between different classifiers. The classification accuracy is a widely-adopted metric used in performance assessment. Due to the good results demonstrated with several works, the classifiers (already introduced in Section 1.3.3 and recalled here for the reader's convenience) typically employed for EEG-based mental state detection are the following ones:

1. *k-Nearest Neighbour (k-NN)* is a non-parametric method that does not need a priori assumption on the data distribution. It uses labeled data for the classification without an actual training phase. Instead, it works as follows: given a set D of labeled points, a distance measure (e.g., Euclidean or Minkowski), and a positive integer k, the k-NN algorithm searches in D for the k points nearest to the new point p, so that the most present class among the k neighbors is assigned to p. Hence, the only *hyperparameters* (i.e, *parameters whose value is used to control the learning process*) required in the k-NN are the positive integer k and the distance measure to use. Eventually, any parameters associated with the chosen distance measure can be considered as a further parameter. These hyperparameters are usually set by using a cross-validation procedure.
2. An *Artificial Neural Network (ANN)* is composed by a set of *neurons* linked to each other into fully connected layers. The output of each neuron is a linear combination of the inputs followed by a non-linear function, which is referred to as *activation function*. The

number of layers, the number of neurons per each layer, and the kind of activation functions are hyperparameters that can be optimized, while the coefficients of linear combination are parameters optimized during the training phase.

3. *Linear Discriminant Analysis (LDA)* is a dimensionality reduction algorithm used to project the features from higher dimension space into a lower dimension space, while trying to preserve the discriminatory information between the classes contained in the data. This method maximizes the separability between data points belonging to different classes. The first step is to calculate the separability between classes (for example, the distance between the means of different classes) i.e., the between-class variance. The second step is to calculate the distance between the mean and data points of each class (the within-class variance). The last step consists in determining the lower dimensional space which maximizes the between-class variance and minimizes the within-class variance.

4. *Support Vector Machines (SVM)* is a binary classifier which classifies input data according to the decision hyperplane having the largest distance from the *margins* of the classes [40]. SVM operates on the basis of a mapping function (*kernel*) which maps each data point into an higher dimensional space. However, the data could not be linearly separable in this space, so non-negative slack variables can be introduced to relax the constraints of the problem by allowing an amount of misclassification.

8

Case Studies

In this chapter, five case studies on mental state detection by passive BCI are presented. Attention in neuromotor rehabilitation is considered in Section 8.1, emotional valence in human-machine interaction in Section 8.2, work-related stress in Section 8.3, and engagement in learning and rehabilitation are presented in Sections 8.4 and 8.5, respectively. In particular, for each of them, data acquisition, signal processing, and results are discussed.

8.1 ATTENTION MONITORING IN NEUROMOTOR REHABILITATION

A system based on an eight-channel EEG with dry electrodes is presented to detect distraction or attention in rehabilitation. The solution allows the detection of distracting conditions, and it enables an automated rehabilitation station to activate strategies to recover attention and maximize rehabilitation efficacy [307]. Indeed, paying attention to rehabilitative exercises increase the effectiveness of the rehabilitation process [170]. Indeed, keeping the attentional focus during the execution of the exercise favors neuronal plasticity and motor learning, i.e, the processes underlying motor recovery [171].

8.1.1 Data Acquisition

The experimental protocol was approved by the ethical committee of the University of Naples Federico II. Seventeen healthy subjects are involved (11 males and 6 females, with an average age of 30.76 $\pm$ 8.15).

Participants are asked to sit on a comfortable chair one meter away from the PC screen and to perform a ball-squeeze exercise every time a start-up command appeared on the PC screen. Ball-squeeze is one of the most common rehabilitation exercises to maintain or restore a functional use of the hand after surgery or in the presence of inflammatory or degenerative diseases [341].

While performing the task, participants have to focus (i) on the task itself or (ii) on a concomitant cognitive distractor. The distraction task is based on the *oddball paradigm* [342, 343]: a sequence of repetitive stimuli interrupted by unusual stimuli. The subjects are required to count the number of a specific

DOI: 10.1201/9781003263876-8

stimuli sequence. Acoustic, visual and acoustic-visual stimuli are presented by using a conventional headphone and a PC screen [344]. Each participant completes a session of 30 trails: 15 in the first and 15 in the second one. The order of trials sequences is random to minimize a task learning effect. In each trial, the stimuli are presented for 2 s, the task execution last 9.5 s and the relaxing phase 5 s. A 15-s baseline is recorded before each session. In particular, in the first condition (*attentive-subject trial*), a ball-squeezing image and a message requiring the subject to focus on the squeezing movement appear on the PC display after a 2-s notification. At the end of the *attentive-subject trial*, an image of a relaxing landscape is presented for 5 s.

In the *distracted-subject trial*, i.e., the second condition, the subject is informed about the type of distractor (audio, visual, or audio-visual) that he/she is going to be introduced. Then, the motor task and the concurrent oddball-based distractor task both begin. The distractor task involve the presentation of sounds at three frequencies: *low* (500 Hz), *middle* (1200 Hz), and *high* (1900 Hz). The probability of occurrence of the first tone is the 50 % while for the middle and high tone is the 25 %. The earphones present eight tones successions. The subject is required to send a notification when a middle tone or a high tone occurs after a low one. In the second condition, a *visual distractor* based on the visual oddball paradigm is employed. Visual stimuli are presented to participants by involving 2D-Gabor masks, namely Gaussian kernel functions modulated with sinusoidal plane wave. Three orientations are considered for three visual stimuli and they are represented in Fig. 8.1: 90°, 60°, and 30°. They have different probability of occurrence: 50 % of probability for the Gabor oriented at 90°, and 25 % of probability for both the Gabor with an orientation of 60° or 30°. Even for this paradigm, eight sequences are presented and the subject is asked to reply when a less probable Gabor mask (i.e., one with orientation equal to 60° or 30°) occurs after the most frequently ones with 90° orientation. Finally, the *audio-visual distractor* task is a com-

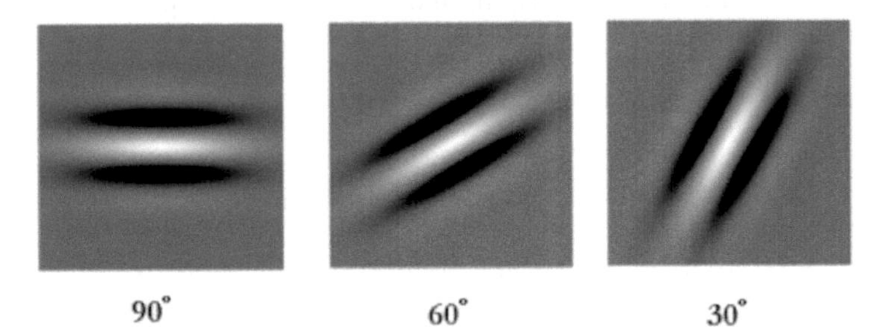

FIGURE 8.1 Visual distractor task elements based on visual Gabor mask with different orientations, equal to 90°, 60°, and 30°, respectively [307].

bination of the previous ones: tones and Gabor sequences appear randomly and the subject is asked to reply only when a visual stimulus occurs after an acoustic one, regardless of the type of acoustic or visual stimulus.

TABLE 8.1
Data-set composition [307].

# Subjects	# Sessions	# Trials per Session	# Epochs per trial	# Epochs per subject	# Total Epochs
17	3	30	3	270	4590

At the end of the session, the subject observes a relaxing landscape for 5 s and he/she is asked for the number of target sequences counted during the session. At the end of the experiments, the collected data consist in 4590 epochs composed of eight channels of 512 samples (see Tab. 8.1 for details). In particular, half of the epochs is acquired during the *attentive-subject trials* and they are labeled as belonging to the first class, while the remaining epochs are acquired during the *distracted-subject trials* and they are labeled as belonging to the second class.

The EEG acquisition system employed for this application is the ab medica Helmate [62]. The signals are acquired with a sampling rate of 512 Sa/s as difference between the eight electrodes and the electrode *Fpz*, with respect to AFz.

8.1.2 Processing

The pre-processing stage first involves several filters: a 12 IIR band-pass Filter Chebyshev type 2 filter bank, 4 Hz amplitude, equally spaced from 0.5 Hz to 48.5 Hz. Then the signals are conditioned by a Texas Instruments Analog Front-End, the ADS1298 [345], with a 24-bit $\Delta - \Sigma$ ADC with built-in Programmable Gain Amplifiers, internal reference, or an on-board oscillator. Features are extracted both in the time and frequency domain. In particular, the spectral features considered are Relative and Absolute Power Spectral Density with reference to the seven traditional EEG bands (delta [1, 4] Hz, theta [4, 8]Hz, alpha [8, 12] Hz, low beta [12, 18]Hz, high beta [18, 25] Hz, low gamma [25, 35] Hz, and high gamma [35, 45] Hz), nine 8 Hz bands with 4 Hz overlap in the range [1, 40] Hz and twelve 4 Hz bands, without overlap, in the range [0.5, 48.5] Hz).

The number of features for each epoch is 112 in the first case, 144 in the second case, and 192 in the third case. Also in the time domain, different solutions are compared, namely features extracted from the CSP application and those extracted from CSP preceded by different types of Filter Banks. The number of features obtained are 8, 56, 72, and 96, respectively.

Five classifiers are compared: kNN [35], SVM [39], ANN [39], LDA [34], and NB [38]. The hyperparameters of a classifier are selected by a random search with a nested CV procedure in order to minimize possible bias due to the reduce sample dimension [346]. Notably, nested CV differs from the classical k-fold CV. In the classic procedure the dataset is divided into a partition of k subsets (folds). Meanwhile, in the nested one, two nested k-fold Cross Validation procedures are employed: an outer one to finds the best hyperparameters of the model, while the inner one to estimate the performance of previous one. Data are separated into training and the test sets. Table 8.2 reports the hyperparameters variation range considered in the present case study.

TABLE 8.2
Hyperparameters considered in classifier optimization and their variation ranges [307].

Classifier	Hyperparameter	Variation Range
k-Nearest Neighbour (k-NN)	Distance (DD)	{cityblock, chebychev, correlation, cosine, euclidean, hamming, jaccard, mahalanobis, minkowski, seuclidean, spearman}
	DistanceWeight (DW)	{equal, inverse, squaredinverse}
	Exponent (E)	[0.5, 3]
	NumNeighbors (NN)	[1, 5]
Support Vector Machine (SVM)	BoxConstraint (BC)	log-scaled in the range [1e-3,1e3]
	KernelFunction (KF)	{gaussian, linear, polynomial}
	KernelScale (KS)	log-scaled in the range [1e-3,1e3]
	PolynomialOrder (PO)	{1,2,3,4}
Artificial Neural Network (ANN)	Activation Function (AF)	{relu, sigmoid, tanh}
	Hidden Layer nr. of Neurons (HLN)	[25, 200]
Linear Discriminant Analysis (LDA)	Gamma (G)	[0,1]
	Delta (D)	log-scaled in the range [1e-6,1e3]
	DiscrimType (DT)	{linear, quadratic, diagLinear,} {diagQuadratic, pseudoLinear, pseudoQuadratic}
Naive Bayes (NB)	DistributionName (DN)	{normal, kernel}
	Width (W)	log-scaled in the range [1e-4,1e14]
	Kernel (K)	{normal, box, epanechnikov, triangle}

8.1.3 Results and Discussion

All the features-classifiers combinations are tested in a within-subjects approach and the better accuracy is achieved with features extracted from the time domain and a 12-filter bank and CSP algorithm. In particular, the application of 12-bands filter bank allows to obtain the best performances with reference to all classifiers except for the LDA. The best accuracy ($92.8 \pm 1.6\%$) is reached with the k-NN. In the specific application, the error minimization during distraction assessment is fundamental. Therefore, the *precision* (true positives and relevant elements ratio) is considered to appreciate the impact of false negatives for distraction class. The results are reported in Fig. 8.2 and they show a mean recall associated with the k-NN higher than 92 %.

The method for distraction detection is particularly suitable for implementation in the rehabilitation field. Thanks to the use of dry electrodes and wireless transmission, a high wearability is possible. This system can be used by the therapists to estimate the attention of a patient during the proposed rehabilitation exercises and, consequently, implement strategies to increase the patient's attention, thus implying an increase therapy efficacy.

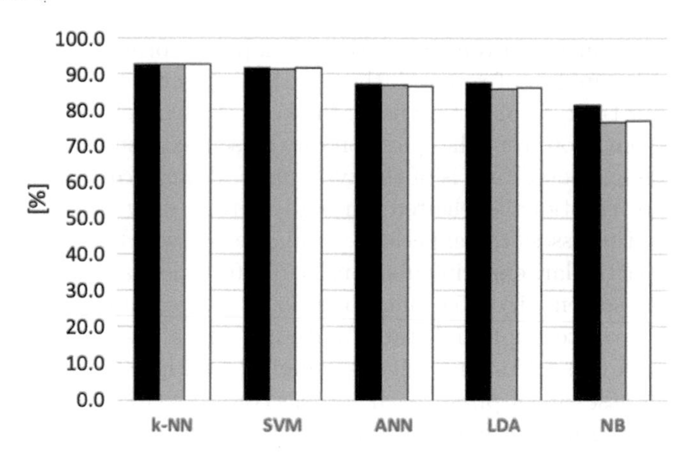

FIGURE 8.2 F-Measure test results for the best performance of each classifier: Precision (black) , Recall (gray), and F1-score (white).

8.2 EMOTION DETECTION FOR NEURO-MARKETING

Neuro-marketing is a branch of neuroscience aimed to study people's brain processes and their changes during decision-making in order to predict their behavior as consumers. Neuro-marketing uses brain imaging techniques to map emotions, and to identify the areas of the brain involved in the process of buying a product or selecting a brand. In particular, the goal of neuro-marketing is to capture consumers' real-time decision making and to highlight processes that cannot be detected by traditional survey techniques based on ex-post questionnaires. This sector is growing strongly: a *compound annual growth rate (CAGR)* of 12 % has been recorded over the last four years. [347] Clearly, in this context, emotions play a key role.

In this section, an EEG-based method to assess emotional valence is presented as case study. In this method, the wearability is guaranteed by a wireless cap with eight conductive-rubber dry electrodes [276].

8.2.1 Data Acquisition

The experimental protocol was accepted by the ethics committee of the Federico II University and subjects are asked for informed consent to publish their personal data and images. Thirty-one subjects are firstly involved and the Patient Health Questionnaire (PHQ) is administered. According to that questionnaire, six participants with depressive disorders are excluded. The experiments are conducted on 25 subjects (52 % male, 48 % female, age 38 $\pm$ 14) and consist in the presentation of emotion-inducting stimuli, based on the Mood Induction Procedure (MIP) [348]. The procedure is conducted in a dark and soundproof room. Although the subjects are not explicitly instructed about the type of stimuli used, they can easily infer the purpose of the stimulus presentation from the question on the self-assessment questionnaire. Therefore, the proposed task is of the type *implicit-more controlled* [349]. The volunteers are emotionally elicited through passive viewing of pictures and they are asked to assess the experienced valence by two classes: negative and positive. In particular, experiments consist of 26 trials, each lasted 30 s: (i) a 5 s-long white screen, (ii) a 5 s-long countdown frame employed to relax the subject and separate emotional states mutually, (iii) a 5 s-long eliciting image projection, and (iv) a 15 s-long self-assessment (see Fig. 8.3 for details).

In the self-assessment phase, each subject compiles the SAM questionnaire by self-evaluating his/her emotional valence (positive or negative) on a scale from 1 to 5. The half of 26 randomly presented images is meant to elicit negative emotion while the other 13, to elicit positive ones.

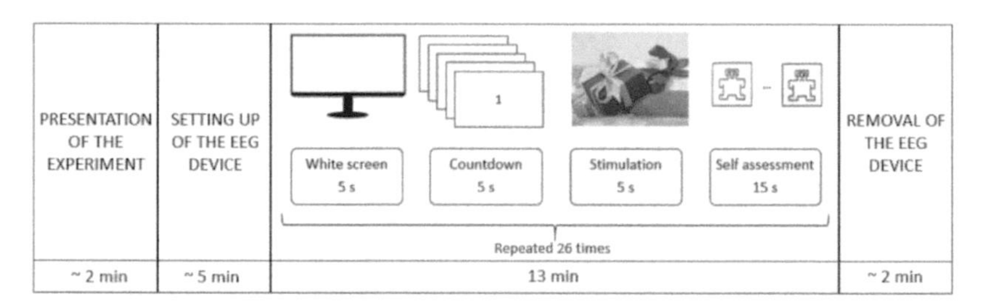

FIGURE 8.3 Experimental protocol [276].

Images with their own score on a scale from 1.00 to 7.00 are selected from the database Oasis [350]. In particular, only images (polarized images) that deviate by one from the maximum and minimum scores (6.28 and 1.32, respectively) of the images of Oasis dataset are selected. In this way, the sensitivity of the measurement system results improved.

Then, a preliminary test is carried out on 12 subjects to exclude relevant discordance between the dataset scores and the scores from self-administered questionnaires. Indeed, sometimes a cultural bias can be found with respect to the emotional elicitation of certain stimuli. An emblematic example are the symbols linked to the Ku Klux Klan. These have a different emotional impact for a citizen of the United States of America than for Europeans. The 12 participants are asked to rate each image using the SAM scale. Results of this preliminary analysis let to exclude images with a neutral rating from at least 50 % of the subjects. Finally, 13 images with a score in the range [1.32, 2.32] and 13 in the range [5.28, 6.28] are randomly selected from the dataset. For these images, in Fig. 8.4, the Oasis valence score and the average scores (across all subjects) of the self-assessment are shown.

The use of 26 images ensures an EEG track of more than 100 s per subject and, at the same time, user comfort, not making the duration of the

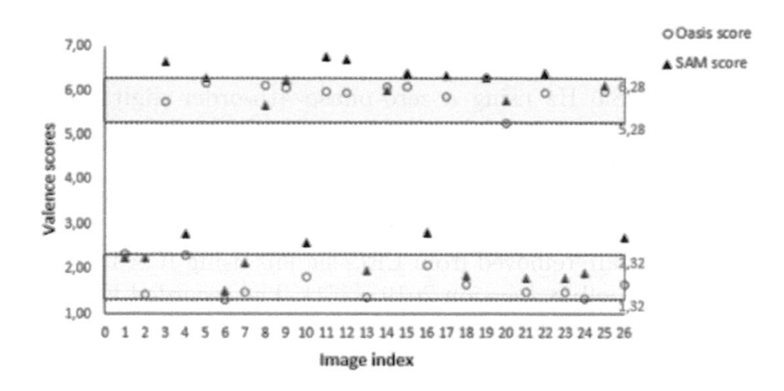

FIGURE 8.4 Oasis valence and SAM average scores of the 26 images selected for the experiments. The Oasis score intervals used to extract polarized images are identified by dotted lines [276].

experiment excessive. In this way, the experiment last about 20 min per subject: 2 min for activity explanation, 5 min for the setting up of the EEG device and 13 min for image presentation). Bland-Altman and Spearman analyses are performed to compare the experimental sample to the Oasis one. The qualitative analysis in Fig. 8.5 and the Spearman correlation index $\rho = 0.799$ show the compatibility between the scores expressed by the two samples. The EEG acquisition system employed for this application is the ab medica Helmate [62].

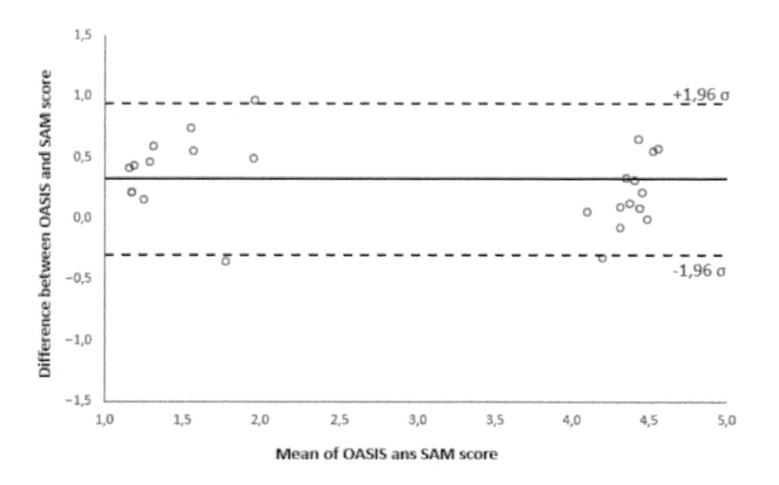

FIGURE 8.5 Bland-Altman analysis on the agreement between stimuli (OASIS) and volunteers perception (SAM) [276].

8.2.2 Processing

The EEG tracks are acquired at a sampling frequency of 512 Sa/s and filtered between 0.5 and 48.5 Hz using a zero-phase 4th-order digital Butterworth filter. In the processing stage, the number of trials used is reduced from 26 to 24 per subject because macroscopic artifacts corrupted two trial per three subjects. Therefore, the number of trials is reduced by removing randomly two trials for all other subjects in order to keep the dataset balanced. The remaining artifacts are removed from EEG signals using ICA by means of the EEGLAB Matlab toolbox (version 2019) [351]. The recorded EEG signals are divided into 2-s windows overlapping of 1 s.

Different approaches of features extraction and classification are compared in order to validate this method (see Section 7.1.2).

As far as the features extraction is concerned, two methods are compared, namely "with" and "without" a priori spatial-frequency knowledge given by neurophysiology. In the first method, frontal asymmetry feature is computed by subtracting the left frontal (*FP1*) from the right (*FP2*) channel by employing a [8, 13] Hz (alpha band) pass-band filter. Indeed, neurophysiological theories confirmed the relevance of the frontal regions in emotional phenomena.

The second method, instead, is based on the application of PCA (by using a number of principal components explaining the 95 % of data variance) or CSP algorithms, for the feature extraction phase. Moreover, the analyses involve either only spatial or the combination of spatial and frequency features, which is obtained by separately computing these features in several frequency bands. In particular, when only spatial information is considered, input features are 8192 (8 channels $\times$ 1024 samples). In case of spatial and frequency combination, features increase up to 98304 (12 frequency bands $\times$ 8 channels $\times$ 1024 samples). The 12 frequency bands result by the application of 12 IIR band-pass filters Chebyshev type 2, with 4 Hz bandwidth, equally spaced from 0.5 Hz to 48.5 Hz. The features are reduced from 98304 to 96 in case of application of the CSP algorithm.

Also in the classification stage, different solutions are compared; as concerns the type of investigation, a within-subject and a cross-subject approach are employed. In the first approach, data from different subjects are analyzed separately while, in the cross-subject method, data from all subjects are merged into one dataset. A 12-fold Cross Validation procedure is used to validate the method. This procedure allows to find the best classifier hyperparameters values. To this aim, data are divided into K subsets (folds). $K-1$ folds are used to train the classifier, while the remaining fold is used to evaluate the model performance. This process is made for all the possible combinations of the K folds. The final result is the average scores on all the test sets. In each step of the aforementioned procedure, the epochs of each trial are kept together to avoid that different part of the same trial can be included in the training and the test sets. In this way, the *N-trial-out* approach prevent

TABLE 8.3
Classifier optimized hyperparameters and variation range [276].

Classifier	Hyperparameter	Variation Range
k-Nearest Neighbour (k-NN)	Distance (DD)	{cityblock, chebychev, correlation, cosine, euclidean, hamming, jaccard, mahalanobis, minkowski,spearman}
	DistanceWeight (DW)	{equal, inverse, squaredinverse}
	Exponent (E)	[0.5, 3]
	NumNeighbors (NN)	[1, 5]
Support Vector Machine (SVM)	BoxConstraint (BC)	log-scaled in the range [1e-3,1e3]
	KernelFunction (KF)	{gaussian, linear, polynomial}
	KernelScale (KS)	log-scaled in the range [1e-3,1e3]
	PolynomialOrder (PO)	{2,3,4}
Artificial Neural Network (ANN)	Activation Function (AF)	{relu, sigmoid, tanh}
	Hidden Layer nr. of Neurons (HLN)	[25, 200]
Linear Discriminant Analysis (LDA)	Gamma (G)	[0,1]
	Delta (D)	log-scaled in the range [1e-6,1e3]
	DiscrimType (DT)	{linear, quadratic, diagLinear, diagQuadratic, pseudoLinear, pseudoQuadratic}
Random Forest (RF)	Depth (D)	[5,20]
	Number of Trees (NT)	[15,100]
	Maximum Depth of the tree	[5,30]
Logistic Regression (LR)	Penalty (P)	{L2, elastic net}
	Inverse of regularization strength (C)	[0.25, 1.0]

to facilitate the classification goal, due to the similarity of data belonging to the train and the test set. In particular, in the within-subject setup, 88 epochs for training and 8 epochs for testing are used, while, in the cross-subject case 2200 and 200 epochs are used in training and test set, respectively.

kNN [352] and ANN [39] are compared with other four classifiers: LDA [292], SVM [353], *Logistic Regression (LR)* [39] and RF [354]. LR is a classification method that works by estimating the probability of a sample of belonging to the $y = 1$ by computing the formula:

$$P(y|\mathbf{x}) = \frac{\exp(q + \mathbf{wx})}{1 + \exp(q + \mathbf{wx})} \tag{8.1}$$

where $\mathbf{w}$ and q are parameters to be identified by learning and $\mathbf{x}$ is the sample. The hyperparameters resulting from the CV procedure are reported in Tab. 8.3 for each classifier. The metric performance considered to compare the results are accuracy, precision, and recall.

8.2.3 Results and Discussion

Results are shown in terms of mean accuracy and standard deviation at varying the adopted classifier in the within-subject case and in cross-subject case. In particular, Tab. 8.4 is referred to a priori spatial-frequency knowledge-EEG features while Tabs 8.5 and 8.6 in reference to no a priori spatial knowledge case.

In both cases (within-subject and cross-subject), better results are achieved without a-priori knowledge and when features are extracted by FBCSP. In this case, classifiers are more accurate in within-subject analysis rather than cross-subject analysis. In the latter data, variability is higher having combined the data of all the subjects. The most performing classifier was k-NN ($k = 2$) in the cross-subject approach, reaching 80.2 % of accuracy while in the within-subject case, ANN (one layer, less than 100 neurons

TABLE 8.4 Percentage accuracy (mean and standard deviation) considering a priori knowledge i.e., Asymmetry—Within-subject (Within) & Cross-subject (Cross) [276].

Classifier	Entire EEG Band		α Band	
	Within	Cross	Within	Cross
k-NN	54.0±4.1	51.0±1.2	53.8±4.0	51.3±0.4
SVM	56.8±3.4	50.8±0.2	56.7±3.0	51.2±0.3
LDA	54.5±3.8	51.2±0.8	53.8±3.5	51.0±1.0
ANN	58.3±3.0	51.8±0.3	58.5±3.0	51.5±1.6
RF	55.7±3.9	50.7± 1.2	54.5± 4.5	50.9±1.3
LR	52.5± 4.1	51.4± 0.2	53.7± 4.3	51.2± 0.7

TABLE 8.5
Percentage accuracy (mean and standard deviation) without considering a priori knowledge i.e., Asymmetry—Within-subject [276].

Classifier	Entire EEG Band			Filter Bank		
	No PCA/CSP	PCA	CSP	No PCA/CSP	PCA	CSP
k-NN	71.0±6.0	67.7±8.4	72.0±8.9	75.6±5.8	66.8±7.2	94.5±3.5
SVM	66.9±8.1	66.3±10.3	73.4±9.5	71.6±8.9	62.0±7.8	95.5±2.8
LDA	63.1±4.9	55.3±4.0	74.0±10.0	62.9±5.3	53.9±3.5	95.0±2.9
ANN	69.7±5.1	66.3±6.2	78.1±8.0	66.7±4.9	65.6± 5.6	**96.1±3.0**
RF	66.4 ± 4.1	58.9 ± 4.2	72.8 ± 9.4	67.4 ± 4.1	59.3 ± 5.0	94.2 ± 2.7
LR	62.7 ± 4.9	52.3 ± 2.9	72.6 ± 9.3	61.0 ± 5.0	51.2 ± 4.0	95.1 ± 2.9

TABLE 8.6
Percentage accuracy (mean and standard deviation) without considering a priori knowledge i.e., Asymmetry—Cross-subject [276].

Classifier	Entire EEG Band			Filter Bank		
	No PCA/CSP	PCA	CSP	No PCA/CSP	PCA	CSP
k-NN	68.4±0.2	62.1±0.9	56.8±0.5	70.1±1.0	61.1 ± 0.3	**80.2±2.1**
SVM	51.5±0.6	52.1±0.3	61.0±2.0	51.8±1.0	51.2±0.3	71.3±2.0
LDA	53.5±0.7	50.9±0.4	55.4±4.2	52.6±0.1	50.9± 0.2	63.7±2.1
ANN	59.9±1.0	54.5±0.2	58.1±1.1	57.4±0.1	53.7±0.1	63.3±2.7
RF	56.5 ± 0.6	55.3 ± 0.7	59.2 ± 1.9	57.8 ± 1.1	52.5 ± 2.9	65.0 ± 3.8
LR	50.5 ± 1.9	50.6 ± 0.5	55.7 ± 4.9	51.8 ± 0.9	50.9 ± 0.5	58.1 ± 1.5

and *tanh* activation function) reached the best accuracy (96.1 %). In Fig. 8.6 precision, Recall and F1-score metrics are reported.

In conclusion, the present case study confirms the compatibility of the experimental sample with that of *Oasis* trough results from the *Self Assessment*

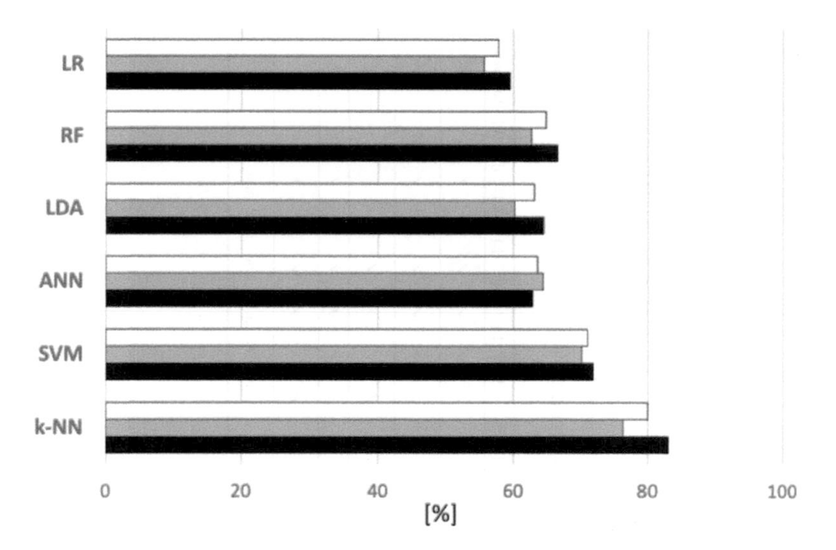

FIGURE 8.6 F1—score (White), Recall (Gray), and Precision (Black) for the best performance of each classifier - Cross-subject [276].

Manikin questionnaire. The combined use of the standardized stimulus and the subject's self-assessment improves the reproducibility of the experiment and the generalizability of the outcome.

The measurement of two emotional valence states (positive and negative valence) along the interval scale theorized by the Circumplex Model of emotions is presented. A comparison between this solution and previous proposals from literature is shown in Tab. 8.7, taking into account the following criteria: (i) classification vs measurement, (ii) standardized stimuli, (iii) self-assessment questionnaires, (iv) number of channels ≤ 10, (v) cross-subject accuracy $> 80\,\%$, (vi) within-subject accuracy $> 90\,\%$. The classification vs measurement issue depends on the reference theory adopted (i.e., Discrete or Circumplex Model). In the literature, only a few studies combine self-assessment questionnaires and standardized stimuli to construct a metrological reference. Therefore, in many cases the generalizability of the results is penalized [272, 274]. Moreover, considering the above criteria, only this method meets them all.

8.3 STRESS ASSESSMENT IN HUMAN-ROBOT INTERACTION

A highly-wearable single-channel instrument, composed of off-the-shelf components and dry electrodes, is illustrated for detecting work-related stress in real time by means of EEG [140].

TABLE 8.7 Studies on emotion recognition classified according to metrological approach, number of channels, and accuracy (n.a. = "not available", ✓ = "the property is verified". "Only for the first line", ✓ = "Measurement") [276].

	[244]	[245]	[238]	[246]	[247]	[239]	[240]	[248]	[249]	[250]	[266]	[251]	[252]	[241]	[253]	[263]	[254]	[242]	[243]	[267]	[268]	[269]	[270]	[271]	[264]	[272]	[273]	[274]	[275]	**Our work**
Measurement vs Classification	✓	✓	✓	✓	✗	✓	✗	✓	✓	✓	✗	✓	✓	✓	✗	✗	✓	✓	✓	✓	✓	✗	✗	✓	✓	✓	✗	✓	✓	✓
Standardized Stimuli	✗	✗	✗	✗	✗	✗	✗	✗	✗	✗	✗	✓	✓	✗	✗	✗	✗	✗	✓	✗	✓	✗	✗	✗	✓	✓	✗	✓	✓	✓
Self-assessment Questionaries	✓	✓	✓	✓	✓	✓	✓	✓	✓	✓	✓	✓	✓	✓	✓	✓	✓	✓	✓	✓	✓	✓	✓	✓	✗	✓	n.a.	✓	✓	✓
#channels≤ 10	✗	✗	✗	✗	✗	✗	✗	✗	✗	✗	✓	✗	✗	✗	✗	✗	✗	✗	✗	✓	✓	✓	✓	✓	✗	✓	✓	✓	✓	✓
Cross-subject Accuracy (>80 %)	✗	✓	✓	✓	✓	✓	✓	n.a	✓	✓	n.a.	n.a	n.a	n.a	n.a	n.a	n.a	✓	✓	n.a	✓	n.a	✓	✗	✗	✓	✗	✗	✗	✓
Within-subject Accuracy (>90 %)	n.a.	n.a.	✓	n.a.	n.a.	n.a.	n.a.	✗	n.a.	n.a.	✓	✓	✓	✓	✗	✓	✗	n.a.	✓	✗	n.a.	✓	n.a.	n.a.	n.a.	n.a.	n.a.	n.a.	✓	✓

In the current fast-changing industrial world, technological innovation, indeed, has introduced new sources of stress. Intelligent automated systems in their various configurations, robots or cobots in collaborative meaning [355], interact continuously with individuals in a relationship of cooperation and, at the same time, of unconscious competition. Robots are strong competitors to humans in several respects, such as memory, computing power, physical strength, etc. Therefore, stress detection in the context of human-robot interaction can help prevent a dysfunctional human-robot relationship.

8.3.1 Data Acquisition

The experiment involves 17 subjects at first. After a questionnaire-based screening, seven subjects are excluded for excessive smoking, high anxiety, and depression scores, as well as low performance on short-term memory tests. Then, the remaining 10 young healthy volunteers (average age 25 years, five women and five men) are divided between control and experimental group and they signed informed consent. A psychologist presents the protocol to the participants. They are asked to perform a *Stroop Color and Word Test (SCWT)* [356], which is a neuropsychological cognitive test widely used in clinical and experimental practice. The test induces a mental load for both the experimental and control group. A negative social feedback is administrated only to the experimental group. Subjects are presented with color-related words printed with a different colored ink than the meaning of the words. They are asked to read the color of ink rather than the word in a limited time and they are informed about eventual errors during the task. The exercise is designed so that it induces stress in the only experimental group by presenting them with complex exercises in a short time and under the effect of emotional stressors. Two self-assessment tests are completed by the participants, both before and after the Stroop Test: the *State-Trait Anxiety Inventory (STAI)* [357] to evaluate current anxiety state, and the *Self-Esteem Inventory* [358] to assess participants' self-esteem. At the end of the tasks, participants also are asked to evaluate the pleasantness of the experience on a Likert scale of 1 to 7.

Each test last 180 s and only the most significant 20 s of an EEG signal are selected for further processing by dividing them into 2 s-long epochs (512 sample per epoch). Ultimately, 20 s of EEG recording are collected for each of the 10 subjects. Therefore, 100 s for both the control and experimental groups are considered. For evaluating the general stress induced to participants, a stress index is considered. This is obtained as the sum of normalized indexes considering performance, anxiety, self-esteem, perceived stress, and motivation, which are obtained from the parametric STAI, Rosemberg tests, and task performance.

One-way ANOVA is perfomed on stress and motivation indexes with a significance level α=0.05. A significant difference emerges between the experimental and the control groups in stress index (F=7.49, p=0.026) and in motivation index (F=14.52, p=0.005): the experimental group is more stressed and less motivated of the control one (see Tab. 8.8 for details on stress index). EEG signals are recorded by two dry active electrodes (Olimex EEG-AE).

TABLE 8.8 Stress index distribution (descending sort) [140].

Subject	Stress Index	Group
1	1.68	Experimental
2	1.66	Experimental
3	1.52	Experimental
4	1.38	Control
5	1.21	Experimental
6	0.77	Experimental
7	0.69	Control
8	0.54	Control
9	0.14	Control
10	-0.06	Control

8.3.2 Processing

Principal Component Analysis (PCA), i.e., a Machine Learning technique to obtain data dimension reduction [330], is applied to data set vectors for the pre-processing stage. A preceding analysis showed that linear classifiers can be used to discriminate data adequately.

The *Leave-p-out cross-validation (LPOCV)* evaluation is conducted in order to generalize the model to new subjects. This procedure improves the statistical significance for small dataset [359], with respect to *Leave-one-out cross-validation (LOOCV)*. The cross-validation is repeated for 10 times in the LOOCV case, and C_p^n times in LPOCV case, where C_p^n is the binomial coefficient, n is the number of subjects in the original sample, and p is the number of subjects reserved only for the test. In the experiment, a leave-two-out is applied (p=2). Keeping training and test set balanced with respect to the two classes (experimental vs control) allow to reach a higher statistical significance (k = 25). In each step, data from one subject for group are used in test set.

Recorded EEG signals are provided as input to four classifiers to discriminate between stressed and unstressed subjects: (i) SVM with linear Kernel, (ii) k-NN with $k = 9$, (iii) RF (criterion = 'gini', max_depth = 118, min_samples_split = 49) , and (iv) ANN (one hidden layer, activation function for hidden node = hyperbolic tangent, loss function = cross entropy cost, post processing = soft max, training algorithm = Resilient Propagation). Optimized hyperparameters of classifier are shown in Tab. 8.9.

TABLE 8.9 Classifier optimized hyperparameters and range of variation [140].

Classifier	Hyperparameters	Variation Range
SVM	Cost parameter (C)	[0.1, 10.1] step = 1.0
Random Forest	n_estimators	{90, 180, 270, 360, 450}
k-NN	n_neighbors	[5, 15] step = 2
ANN	number of internal node	{25, 50, 100, 200}

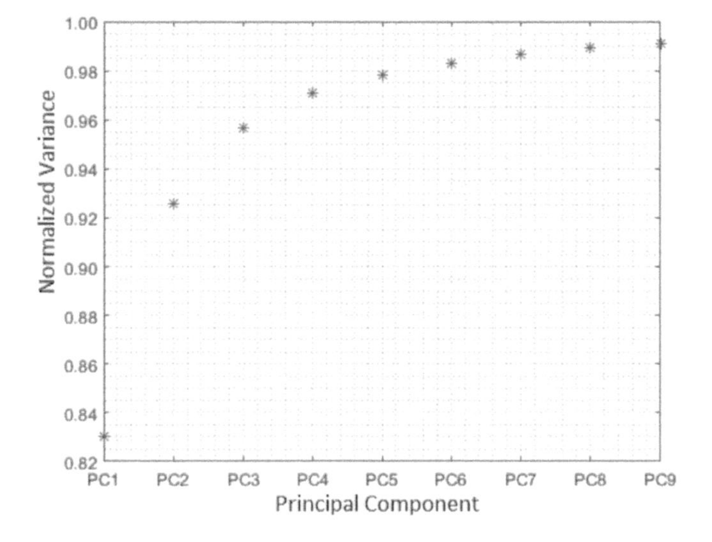

FIGURE 8.7 Cumulative Explained Variance in the PCA [140].

Raw data and PCA pre-processed data are given in input to the classifiers. Each classifier is fed with both raw data (2 s-long EEG epoch) and PCA pre-processed data. The p principal components are computed only for the training set at each iteration of the LPOCV method [359]. The number of principal components varies in range $p \in \{0, 1, 2, \ldots, 9\}$. Results are reported in Tab. 8.10. When PCA is applied, the 99 % of data variance is explained by the first nine components (Fig. 8.7). This demonstrate that intrinsic data dimensionality is a maximum of nine.

8.3.3 Results and Discussion

Results, in terms of mean and uncertainty, are shown in Tab. 8.11. The lower performance is reached by SVM relating to raw data and F-measure was computed to assess SVM performance (Tab. 8.12).

TABLE 8.10 Accuracy (mean and uncertainty percentage) in Original Data (O.D.) and Principal Components Hyperplanes [140].

		Mean Value ± Uncertainty (%)
O.D. Hyperplane		97.5 ± 0.6
	PC1	90.5 ± 5.3
	PC2	78.5 ± 7.1
	PC3	93.2 ± 3.4
	PC4	98.3 ± 0.4
P.C. Hyperplanes	PC5	97.8 ± 0.4
	PC6	97.4 ± 0.6
	PC7	97.8 ± 0.5
	PC8	97.4 ± 0.6
	PC9	97.9 ± 0.5

TABLE 8.11 Classifiers accuracy (mean and uncertainty percentage) in Original Data (O.D.) and Principal Components Hyperplanes [140].

	SVM	**Random Forest**	**k-NN**	**ANN**
O.D.	97.5 ± 0.6	98.5 ± 0.3	98.5 ± 0.4	99.2 ± 3.1
PC1	90.5 ± 5.3	98.6 ± 0.3	98.9 ± 0.3	98.5 ± 3.8
PC2	78.5 ± 7.1	98.8 ± 0.2	98.0 ± 0.5	98.8 ± 3.9
PC3	93.2 ± 3.4	98,4 ± 0.5	98,5 ± 0.4	98.7 ± 4.3
PC4	98.3 ± 0.4	98.9 ± 0.3	98.5 ± 0.4	99.1 ± 2.4
PC5	97.8 ± 0.4	98.8 ± 0.5	98.5 ± 0.4	99.2 ± 2.8
PC6	97.4 ± 0.6	98.4 ± 0.5	98.5 ± 0.4	98.9 ± 3.3
PC7	97.8 ± 0.5	99.0 ± 0.4	98.5 ± 0.4	98.9 ± 3.6
PC8	97.4 ± 0.6	98.6 ± 0.5	98.5 ± 0.4	99.0 ± 3.5
PC9	97.9 ± 0.5	98.9 ± 0.5	98.5 ± 0.4	98.9 ± 4.1

SVM classification output when p=2 is shown in Fig. 8.8 in PCs space. The PCs plot shows vectors distribution with respect to Support Vector. In that, the diamonds are associated to the control group, while the circles represent the experimental group.

Results show a good discrimination of data related to stress state. Good results are achieved also by employing only the first principal component explaining almost the 90 % of variance.

In order to assess the robustness of this method, two tests are employed adding two different kinds of noise. The first is a random Gaussian noise with zero-mean and $\sigma \in \{0.04, 0.08, 0.12, 0.16, 0.20\}$, multiplied by the absolute value of the data maximum. The SVM is used for classification. Results are reported in Tab. 8.13. The second test is employed to verify the occurrence of possible bias during acquisition and consists in adding a random constant value to EEG signal of the test set. Results are reported in Tab. 8.14.

TABLE 8.12 F-measure test results for SVM (mean and uncertainty percentage) [140].

		Precision (%)	Recall (%)
O.D. Hyperplane		96.5 ± 1.0	98.4 ± 0.7
P.C. Hyperplanes	PC1	89.2 ± 5.1	92.2 ± 5.3
	PC2	81.1 ± 6.2	81.1 ± 6.7
	PC3	96.4 ± 1.5	93.6 ± 3.1
	PC4	98.2 ± 0.5	98.5 ± 0.7
	PC5	97.2 ± 0.5	98.5 ± 0.7
	PC6	96.4 ± 1.1	98.5 ± 0.7
	PC7	97.2 ± 0.8	98.5 ± 0.7
	PC8	96.4 ± 1.1	98.5 ± 0.7
	PC9	97.2 ± 0.8	98.7 ± 0.6

TABLE 8.13 Accuracy (mean and uncertainty%) in Original Data (O.D.) and Principal Components Hyperplanes at varying amplitude of random Gaussian noise [140].

	Noise σ Percentage Value				
	4	**8**	**12**	**16**	**20**
O. D.	97.9 ± 0.5	97.0 ± 0.7	96.1 ± 0.8	95.2 ± 0.9	92.2 ± 1.2
PC1	90.1 ± 5.3	89.7 ± 5.2	88.2 ± 5.2	86.4 ± 5.1	84.1 ± 4.8
PC2	79.0 ± 7.0	77.9 ± 7.1	77.5 ± 6.7	74.7 ± 6.6	74.5 ± 6.5
PC3	93.2 ± 3.4	92.4 ± 3.5	88.7 ± 3.4	85.7 ± 3.3	84.4 ± 3.4
PC4	98.2 ± 0.4	97.1 ± 0.7	95.9 ± 0.7	92.7 ± 0.9	90.8 ± 1.1
PC5	97.6 ± 0.4	97.2 ± 0.5	94.6 ± 0.7	91.6 ± 0.1	89.9 ± 0.1
PC6	97.7 ± 0.6	96.6 ± 0.7	95.1 ± 1.0	93.3 ± 1.0	90.4 ± 1.1
PC7	97.9 ± 0.5	96.8 ± 0.7	96.2 ± 0.8	93.7 ± 0.9	91.6 ± 0.1
PC8	97.5 ± 0.6	96.8 ± 0.6	96.3 ± 0.7	93.5 ± 0.1	90.5 ± 0.1
PC9	98.1 ± 0.5	97.2 ± 0.6	96.6 ± 0.6	93.7 ± 0.9	91.9 ± 0.1

In both noise and robustness tests, good results were achieved but performances degraded with higher noise levels. In particular, performance deteriorated when the noise level is greater than 12 % of absolute value of the maximum of data.

8.4 ENGAGEMENT DETECTION IN LEARNING

An EEG-based method to assess engagement state during learning activity is presented in this case study. Engagement is analyzed as a combination of

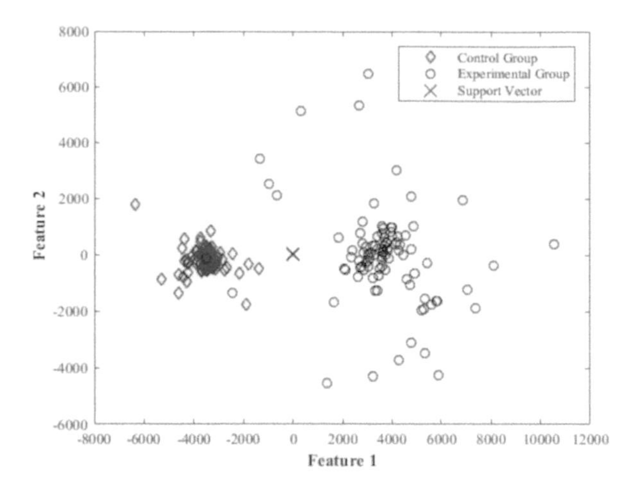

FIGURE 8.8 SVM data distribution in PCs space, p = 2, 92.6 % of explained variance [140].

TABLE 8.14 Accuracy (mean and uncertainty) in Original Data (O.D.) and Principal Components Hyperplanes at varying amplitude of homogeneous noise (%)

	Noise Percentage Value				
	4	**8**	**12**	**16**	**20**
O. D.	**97.4 ± 0.6**	**97.4 ± 0.6**	**97.1 ± 0.7**	**92.6 ± 2.2**	**88.7 ± 3.3**
PC1	90.7 ± 5.3	90.8 ± 5.2	90.5 ± 5.3	89.9 ± 5.4	88.0 ± 5.6
PC2	78.0 ± 7.2	77.4 ± 7.2	76.0 ± 7.4	73.4 ± 7.2	68.9 ± 7.0
PC3	93.0 ± 3.6	90.8 ± 3.8	85.1 ± 4.5	81.8 ± 4.5	72.0 ± 4.5
PC4	98.2 ± 0.4	97.4 ± 0.7	94.2 ± 1.8	89.7 ± 3.2	85.2 ± 3.2
PC5	97.5 ± 0.4	97.6 ± 0.4	93.5 ± 1.8	88.8 ± 3.1	85.3 ± 3.3
PC6	97.4± 0.6	97.8 ± 0.5	95.6 ± 0.1	91.1 ± 3.2	87.6 ± 3.4
PC7	97.8 ± 0.5	97.8 ± 0.5	95.4 ± 1.5	91.1 ± 3.1	87.6 ± 3.4
PC8	97.5 ± 0.6	97.4± 0.6	94.9 ± 1.5	89.9 ± 3.0	86.7 ± 3.4
PC9	97.8 ± 0.5	97.5 ± 0.5	94.9 ± 1.54	90.9 ± 2.9	88.1 ±3.3

emotional and cognitive dimension. The use of a low number of dry electrodes makes the system highly wearable and suitable for usage in the school environment [327]. Learning processes consists in building new mental schemes. When the subject learns new skills and/or modifies already learned ones, neuronal maps are created and/or enriched. Being engaged in the learning process fosters this mechanism. In particular, the evaluation of cognitive and emotional engagement during school lessons can led to (i) academic program adaptation to individual needs and (ii) maximization of class involvement through real-time feedback about student engagement to the teacher in real time.

8.4.1 Data Acquisition

Twenty-one subjects without neurological disorder (9 males and 13 females, 23.7 ± 4.1 years) are involved in the experiment. Each subject signs a written consent and the experimental protocol was approved by the ethical committee of the University of Naples Federico II.

Experiments take place in a room with a comfortable chair at a distance of 1 m from the computer screen. Each subject wears the ab medica Helmate [62] and a check on the contact impedance is carried out in order to increase the quality of acquired signals.

Each experimental session is composed of eight trials. In each trial, the subjects complete revised *Continuous Performance Test* (*CPT*) [360] represented in Fig. 8.9. Three types of tasks are presented to the participants. In particular, in the first, the subject is asked to hold a black cross within a red circle by using the PC mouse. The speed of the cross increases with each trial. At the end of each trial, the ratio between the time in which the cross is inside the circle and the total time is calculated. In the second task, a background music is randomly selected from the MER [361] database. *Cheerful music* and *sad music* are associated to high and low emotional engagement, respectively, in accordance with the Russell's Circumplex Model of emotions [362]. During the first, third, fourth and fifth trial, the songs are played to stimulated high levels of emotional involvement, while the opposite happens for the other trials. In the last task, different emotive engagement conditions are elicited in the subjects with encouraging or disheartening comments on his/her performance. Simultaneously, background music is chosen to elicit a congruent emotion.

In order to validate the stimuli effectiveness, a 9-point scale version of the SAM is proposed to the participants and an empirical threshold is used to evaluate the CPT stimuli response, adapting it to the difficulty level of the task. At the beginning of each session, a preliminary CPT training is conducted in order to uniform the level of participants. After each CPT trial the SAM is assessed.

At the end of the experimental phase, 45 s of EEG acquisitions are collected for each participant and labeled according to the emotional and cognitive engagement levels (low or high). As far as cognitive engagement is concerned, trials with a speed lower than 150 pixels/s are labeled as low_c while trials having speed higher than 300 pixels/s are labeled as $high_c$. This labeling ensures a separation margin between the two classes of 150 pixels/s. The labeling of emotional engagement is based on the type of background music and social feedback: positive feedback and cheerful are associated to high emotional engagement while sad music and negative feedback to the low one. As aforementioned, for each trial SAMs questionnaires analysis is performed to verify

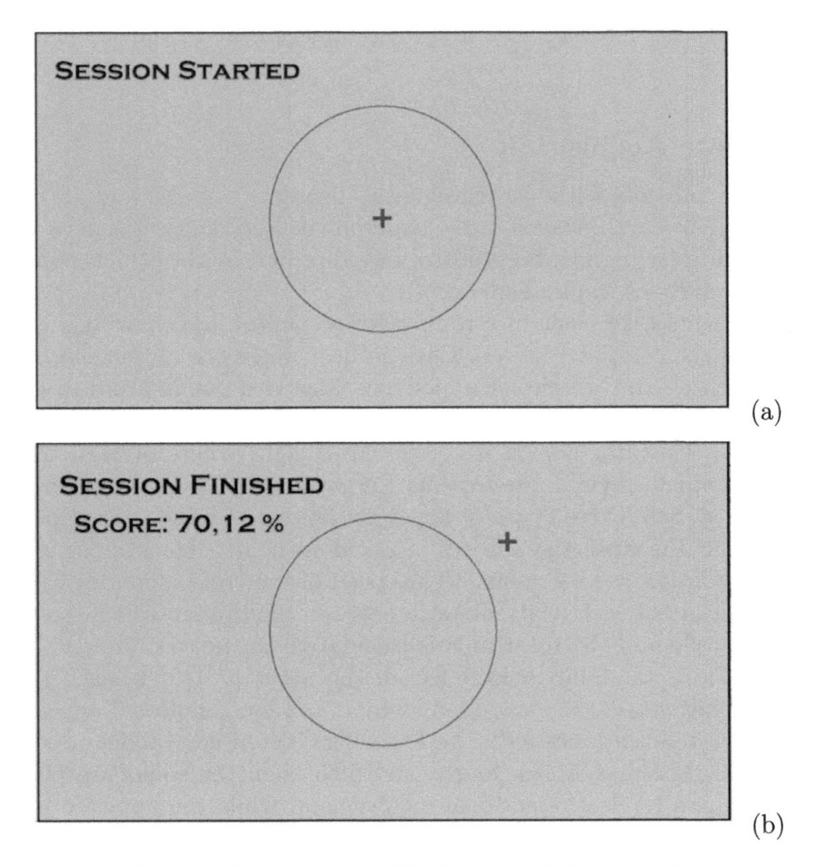

FIGURE 8.9 Screen shots from the CPT game [327]. At the beginning of the game, the cross starts to run away from the center of the black circumference (a). The user goal is to bring the cross back to the center by using the mouse. At the end of each trial, the score indicates the percentage time spent by the cross inside the circumference (b).

the congruence of the elicited response with the offered stimulus. A one-tailed t-student analysis confirmed the congruence (worst case p-value = 0.02). The EEG signals are acquired by Helmate from ab medica [62].

8.4.2 Processing

ICA is employed for artifact removal trough the *Runica* module of the EEGLab tool [363]. After data normalization (subtraction of the mean and division by the standard deviation), the five aforementioned processing strategies are compared (BPCA, BCSP, FBCSP, Domain Adaptation, and Engagement Index computation). In the classification stage, the SVM, k-NN, shallow ANN, and LDA are compared.

TABLE 8.15 Within-subject classification accuracies (%) using the *Engagement Index* [364] for cognitive engagement classifications [327].

Method	Cognitive Engagement
SVM	54.8 ± 4.9
k-NN	53.7 ± 5.7
ANN	53.1 ± 5.4
LDA	50.7 ± 6.2

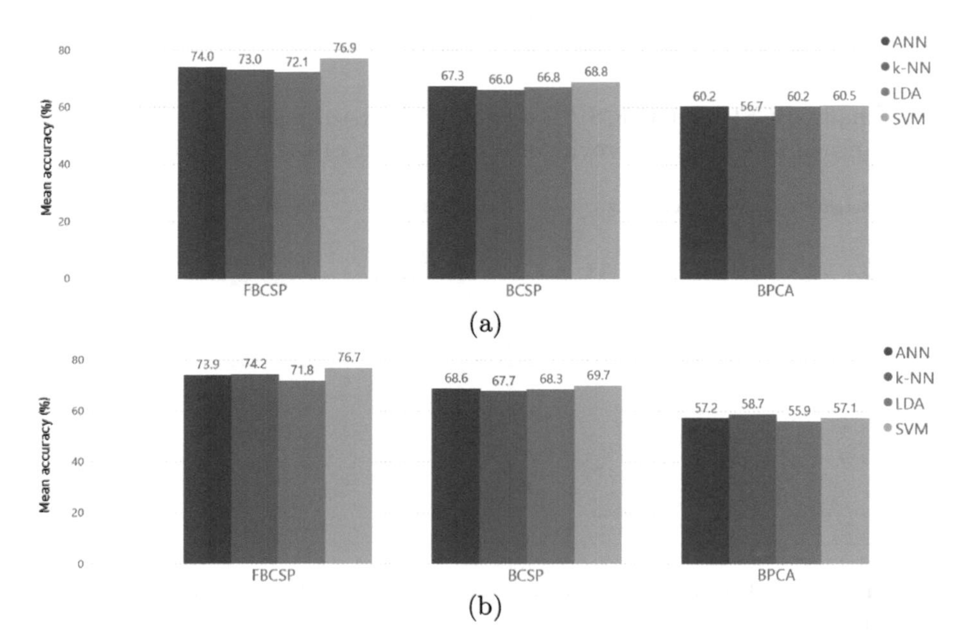

FIGURE 8.10 Within-subject performances of the compared processing techniques in (a) cognitive engagement and (b) emotional engagement detection. Each bar describes the average accuracy over all the subjects [327].

8.4.3 Results and Discussion

A within- and cross-subject approach is carried out. In the first one, a preliminary step consisted in computing the classical engagement index proposed in [364]. This strategy leads to low accuracy performance as shown in Tab. 8.15. The application of the pipeline FBCSP on both cognitive and emotional engagement (Fig. 8.10) allow to achieve good accuracy values. In particular, the SVM reveals a better performance than the other classifiers, reaching an accuracy of 76.9 ± 10.2 on cognitive engagement classification and of 76.7 ± 10.0 on emotional engagement. In Tab. 8.16, results in terms of average accuracy over all the subjects, are reported.

TABLE 8.16 Within-subject experimental results. Accuracies (%) are reported on data pre-processed using Filter Bank and CSP for both cognitive and emotional engagement classifications. The best performance average values are highlighted in bold [327].

Method	Cognitive Engagement (proposed)	Emotional Engagement (proposed)
SVM	**76.9 ± 10.2**	**76.7 ± 10.0**
k-NN	73.0 ± 9.7	74.2 ± 10.3
ANN	74.0 ± 9.2	73.9 ± 9.1
LDA	72.1 ± 11.4	71.6 ± 9.3

Finally, BCSP and FBCSP are compared. Results show the better performance of FBCSP in improving the separability of the data (Fig. 8.11). As

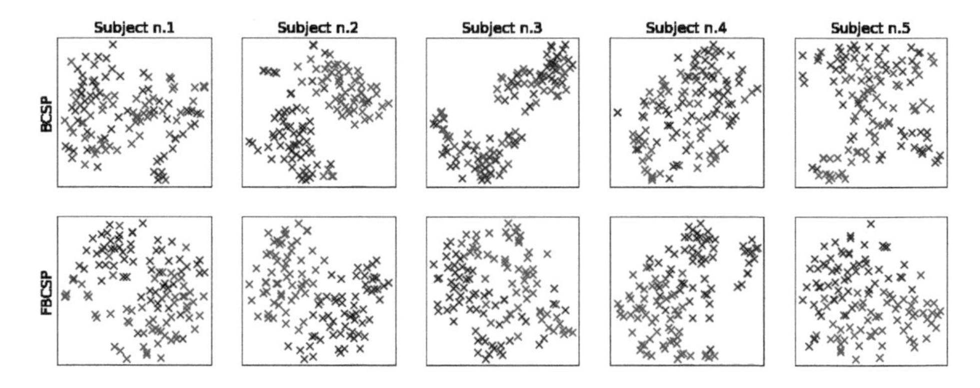

FIGURE 8.11 Filter Bank impact on the class (red and blue points) separability. t-SNE-based features plot of five subjects randomly sampled (first row: without Filter Bank; second row: with Filter Bank) [327].

far as cross-subject approach is concerned, a subject-by-subject average removal is carried out. Data pre- and post-average removal are shown in Fig. 8.12. A greater homogeneity characterized data after the subject-by-subject average removal with respect to the ones without for-subject average removal (Fig. 8.12). In addition, applying TCA after removing the average from each subject allows to obtain better performance with respect to using TCA alone (Tab. 8.17).

In conclusion, the best performances are achieved by using the FBCSP and SVM. In particular, in the cross-subject case, 72.8 % and 66.2 % of average accuracy were achieved by using TCA and for-subject average removal in cognitive and emotional engagement detection, respectively. An accuracy of 76.9 % and 76.7 % was reached in the within-subject case, for the cognitive engagement and emotional engagement, respectively.

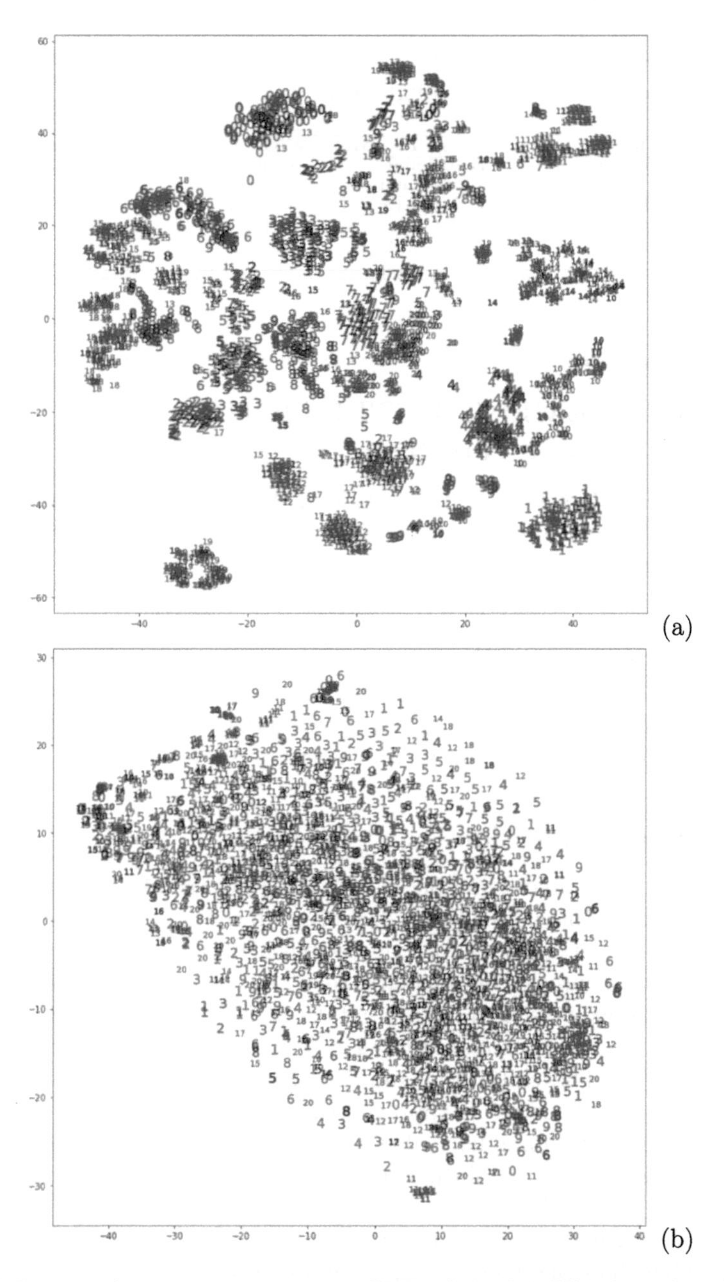

FIGURE 8.12 A comparison using t-SNE of the FBCSP data first (a) and after (b) removing the average value of each subject, in the cross-subject approach [327].

TABLE 8.17 Cross-subject experimental results using FBCSP followed by TCA. Accuracies (%) are reported with and without for-subject average removal for cognitive engagement and emotional engagement detection. The best performance values are highlighted in bold [327].

Method	With For-Subject Average Removal		Without For-Subject Average Removal	
	Cognitive Engagement	Emotional Engagement	Cognitive Engagement	Emotional Engagement
SVM	**72.8 ± 0.11**	**66.2 ± 0.14**	64.0 ± 0.11	61.7 ± 0.10
k-NN	69.6 ± 0.11	61.9 ± 0.09	57.1 ± 0.09	56.9 ± 0.10
ANN	72.6 ± 0.12	65.7 ± 0.14	69.7 ± 0.12	65.8 ± 0.15
LDA	69.5 ± 0.12	65.3 ± 0.14	69.6 ± 0.13	64.6 ± 0.13

8.5 ENGAGEMENT DETECTION IN REHABILITATION

As aforementioned, engagement detection in clinical practice allows to personalize treatments and improve their effectiveness: patients involved in healthcare decision-making tend to perform better. In this case study, an EEG-based method for engagement assessment in pediatric rehabilitation is presented [365]. Cognitive and emotional engagement are considered. In the rehabilitation field, studies generally consider only cognitive engagement and were conducted on adults. Engagement detection in rehabilitation context allows the adaptability of the treatment and a greater effectiveness [317]. The use of portable and wireless EEG cap guarantees high wearability.

8.5.1 Data Acquisition

Four children (three males and one female aged between five and seven years) with neuropsychomotor disorders are involved in the experiment. The children suffer from different diseases such as double hemiplegia, motor skills deficit with dyspraxia, neuropsychomotricity delay, or severe neuropsychomotricity delay in spastic expression from perinatal suffering. Lack of strength, motor awkwardness, difficulty in maintaining balance, inadequate postures are the main symptoms. The children's families signed a written informed consent and the University Federico II approved the experimental protocol. The trials involve children performing normal rehabilitation with simultaneous recording of the EEG signals for 30 min per week for each patient (observational non-interventional protocol).

The experiments take place in a naturally lit room, children are seated on a comfortable chair one meter away from a computer screen (Fig. 8.13). The rehabilitation procedure is carried out according to the *the Perfetti-Puccini method* [366] and consists in performing visual attention tasks. Children are asked to stare at the character presented on the computer screen while keeping the correct position of the neck, head, and trunk. The character could be a bee, a ladybug, a girl, or a little fish. The character, its dimension and direction, the background landscape, and some music can be set by the therapist, so as to make the interactive game more interesting and adapt its difficulty to the patients' needs.

FIGURE 8.13 Neuromotor rehabilitation session [307].

The child's emotional and cognitive involvement states experienced during the exercise are assessed by a multidisciplinary team. The assessment involves the use of visual aids, and the *Pediatric Assessment of Rehabilitation Engagement (PARE)*. In the former case, frontal and lateral videos are taken by two video cameras during the experiment. In the latter case, PARE is used for assessing the emotional, cognitive, and behavioral components of engagement in terms of participation, attention, activation, understanding, positive reactions, interest and enthusiasm, posture, and movements of the child during the exercise. These are quantified on a scale from 0 to 4. Emotional and cognitive engagement are scored as low or high.

The statistical consensus among evaluators in detecting engagement level is of 95.2 % [367]. For each child, sessions of 15 min are employed and a mean of 280 ± 46 epochs for subject is acquired for a total of 1121 epochs, each lasting 9 s (Fig. 8.14). Then, each epoch is labeled based on the engagement state. The 14-channel *Emotiv Epoc+* [63] is used for EEG signal acquisition.

8.5.2 Processing

In the pre-processing stage, the EEG signals are filtered through a 4th order Butterworth band-pass filter between 0.5 Hz and 45 Hz. Then, the artifacts are removed by using ICA. The features extraction involves the application of the CSP whose components optimal number was found through a grid search Cross Validation procedure, varying the number of components in the range [4–14]. The dataset resulted unbalanced because of the non-interventional observational protocol adopted to ensure the comfort of children. Therefore, five

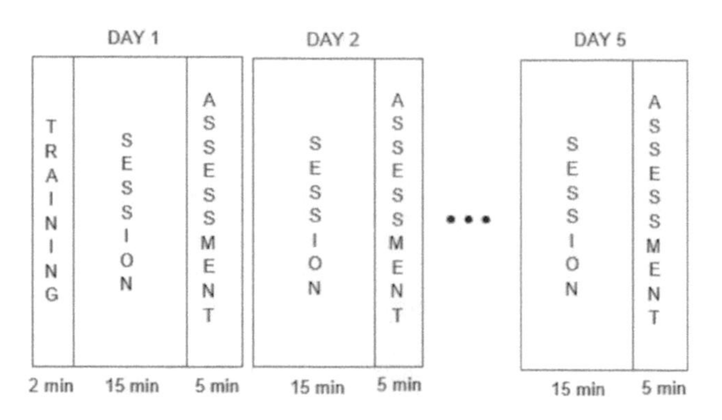

FIGURE 8.14 The experimental paradigm: only on the first day, a training phase is implemented [307].

oversampling methods are employed in order to obtain the same number of data for each class: SMOTE [335], BorderlineSMOTE [336], ADASYN [337], SVMSMOTE [338], and KMeansSMOTE [339]. In Tab 8.18 the hyperparameters values for the five oversampling methods are presented. In the classification stage, three Machine Learning classifiers are used: k-NN, SVM, and ANN.

TABLE 8.18 Oversampling methods, optimized hyperparameters, and variation ranges [307].

Oversampling Method	Optimized Hyperparameter	Variation Range
SMOTE	K Nearest Neighbours	{5, 7, 11}
ADASYN	K Nearest Neighbours	{5, 7, 11}
SVMSMOTE	K Nearest Neighbours	{5, 7, 11}
BorderlineSMOTE	K Nearest Neighbours	{5, 7, 11}
KMeansSMOTE	k-Means Estimator	{1, 2, 4}
	Cluster Balance Threshold	{0.1, 0.9}
	K Nearest Neighbours	{2, 4, 8}

As far as the classification stage is concerned, the hyper-parameters of the three adopted classifiers are found through a grid-search Cross Validation procedure and reported in Tab. 8.19. In particular, SVM require the regularization term C, the kernel type (polynomial or radial basis function) and the degree of the polynomial kernel function (in case of polynomial kernel). In ANN, hyperparameters are the number of neurons and hidden layers and the learning rate. Finally, the best number of neighbors and the distance function need to be set in the k-NN. The classifier is trained on the first 70 % of the data of each session and the remaining data are used as test set.

TABLE 8.19
Classifiers, optimized hyperparameters, and variation ranges [307].

Classifier	Optimized hyperparameter	Variation Range
k-Nearest Neighbour (k-NN)	Distance	{minkowski, chebychev, manhattan, cosine, euclidean}
	Distance Weight	{equal, inverse, squaredinverse}
	Num Neighbors	[1, 7] step: 1
Support Vector Machine (SVM)	C Regularization	{0.1, 1, 5}
	Kernel Function	{radial basis, polynomial}
	Polynomial Order	{1, 2, 3}
Artificial Neural Network (ANN)	Activation Function	{relu, tanh}
	Hidden Layer nr. of Neurons	[5, 505] step: 20
	Learning Rate	{0.0005, 0.0001, 0.001, 0.005, 0.01}

8.5.3 Results and Discussion

In Tabs 8.20 and 8.21, results in terms of intra-individual *Balance Accuracy (BA)* and *Matthews Correlation Coefficient (MCC)* scores are reported for both the cognitive and the emotional engagement, respectively. The chosen metrics ensure that the results are not affected by unbalancing bias in the test phase. The classification is carried out with and without the oversampling methods.

TABLE 8.20 Overall mean of the intra-individual performances on cognitive engagement using three different classifiers: the BA (%) and the MCC at varying the oversampling methods [307].

Oversampling	Metric	k-NN	SVM	ANN	Mean
none	BA	67.1	67.4	73.7	69.4 ± 3.0
	MCC	0.31	0.34	0.45	0.36 ± 0.06
SMOTE	BA	68.6	69.8	72.0	70.1 ± 1.4
	MCC	0.33	0.36	0.40	0.36 ± 0.03
BorderlineSMOTE	BA	70.3	70.9	73.6	71.6 ± 1.4
	MCC	0.36	0.38	0.43	0.39 ± 0.03
ADASYN	BA	68.1	68.3	72.5	69.6 ± 2.0
	MCC	0.33	0.33	0.42	0.36 ± 0.04
SVMSMOTE	BA	69.0	69.4	72.9	70.4 ± 1.7
	MCC	0.34	0.36	0.42	0.37 ± 0.03
KMeansSMOTE	BA	69.8	71.1	74.5	**71.8 ± 1.98**
	MCC	0.35	0.39	0.46	**0.39 ± 0.04**

Results show the effective improvement reached with application of the oversampling methods. In particular, the improvements are more significant for emotional engagement (accuracy increased about 10 %) with respect to cognitive engagement. These could be due to the fact that, in the case of emotional engagement, the available data are strongly unbalanced by class and the SMOTE algorithms had a greater impact on the classification performance. Therefore, the number of data for each class of cognitive engagement is equated to that of emotional engagement and the following results are achieved: the BA and MCC were 58.73 % and 0.25, respectively without any oversampling strategy, and 65.14 % and 0.28 with KMeansSmote oversampling. These results confirm the better suitability of oversampling method to very unbalanced

TABLE 8.21 Overall mean of the intra-individual performances on emotional engagement using three different classifiers: the BA (%) and the MCC at varying the oversampling methods [307].

Oversampling	Metric	k-NN	SVM	ANN	Mean
None	BA	56.3	57.0	61.4	58.2 ± 2.2
	MCC	0.16	0.20	0.26	0.21 ± 0.04
SMOTE	BA	57.6	61.2	67.1	62 ± 3.9
	MCC	0.16	0.24	0.35	0.25 ± 0.08
BorderlineSMOTE	BA	57.3	60.2	66.5	61.3 ± 3.8
	MCC	0.15	0.22	0.34	0.24 ± 0.08
ADASYN	BA	57.0	60.0	67.4	61.5 ± 4.4
	MCC	0.15	0.21	0.36	0.24 ± 0.09
SVMSMOTE	BA	57.3	61.0	64.4	60.9 ± 2.9
	MCC	0.15	0.25	0.31	0.24 ± 0.06
KMeansSMOTE	BA	57.9	63.6	71.2	$\mathbf{64.23 \pm 5.4}$
	MCC	0.18	0.30	0.43	$\mathbf{0.30 \pm 0.10}$

data. Best results are achieved by applying KMeansSMOTE method both for cognitive and emotional engagement.

As far as intra-subjective analysis is concerned, results are shown in Figs. 8.15 and 8.16 by considering the KMeansSMOTE oversampling method. Better performances are achieved by the ANN.

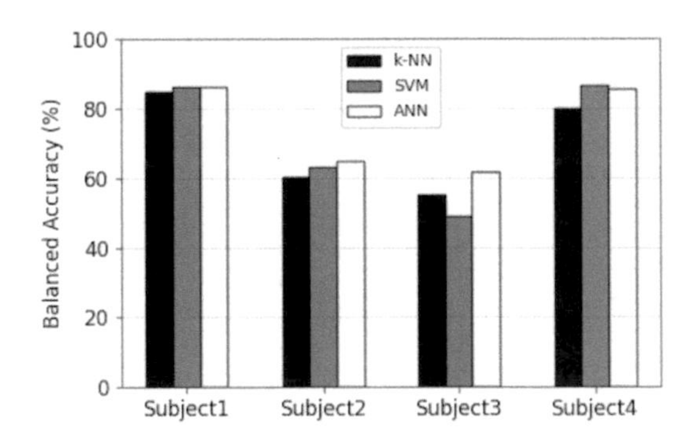

FIGURE 8.15 Cognitive engagement balanced accuracies for each subject based on KMeansSMOTE oversampling technique. Classifier performances are reported [307].

An important consideration must be given concerning the device used for recording the EEG signal: *Emotiv Epoch+* [63] is only partially adaptable to different head sizes. However, in the present experiment, all subjects expose an inion-nasion distance that is within the expected range for the use of such a device ([31.0, 38.0] cm) [368]. In particular, the lower inion-nasion distance is 31.0 cm, therefore, the maximum electrode dislocation is about 1.4 cm with respect to the 10-20 International Positioning System. This information is necessary to make reproducible the measurement.

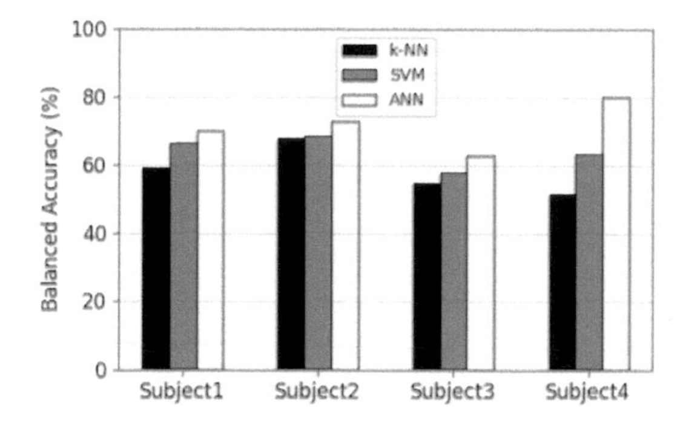

FIGURE 8.16 Emotional engagement balanced accuracies for each subject based on KMeansSMOTE oversampling technique. Classifier performances are reported [307].

The main disadvantage of the implemented device is the low number of electrodes in the parietal region of the scalp. This area is particularly important in spatial attention detection. However, *P7* and *P8* electrodes are sufficient for the presented application where attention is only one component of engagement.

As regards the implications and applications of this method, adaptivity is currently based on performance monitoring in the rehabilitation field. Characteristics not directly observable (such as patient engagement) are usually not taken into consideration. Conversely, the monitoring and the proper stimulation of patient engagement can strongly improve the effectiveness of the rehabilitation intervention. For example, in the framework of neuromotor rehabilitation, maintaining the attention focus on the exercises promotes neuronal neuroplasticity and motor recovery [170]. Therefore, monitoring cognitive engagement allows automated systems to adopt appropriate countermeasures when distraction is detected [307].

Rehabilitation performance is also conditioned by the emotional engagement. A low performance may depend, for example, on a state of boredom or worry, rather than on a lack of skills. Chronic health disabilities are often stressors and the stress management is a crucial issue in rehabilitation [369]. The assessment of cognitive and emotional engagement allows to monitor stress levels [370] and to provide the automated rehabilitation platform useful information to better adapt to the user's needs. Finally, this approach is data driven. Thus, it can be applied flexibly to different targets by identifying *ad-hoc* models suitable for different labeled groups.

Part IV

Active Brain-Computer Interfaces

9

Fundamentals

This chapter introduces active BCI to address wearable instrumentation for control applications. In particular, *Motor Imagery* (*MI*)-based BCIs are discussed. The metrological problems associated with the measurement of MI are first stated in Section 9.1. The background on MI is recalled from a metrological point of view, and then the neurofeedback is introduced as a reference for improving the imagination of a movement. Next, in Section 9.2, the measurement of neurological phenomena related to MI is discussed. This involves essential reasoning on how to detect MI prior to realizing MI-based instrumentation.

9.1 STATEMENT OF THE METROLOGICAL PROBLEM

Active BCIs are very attractive since they rely on spontaneous brain activity, voluntarily generated by a subject, without the need of any external stimulation. However, they typically require long training for both the user and the algorithm before a proper recognition of brain patterns, especially when non-invasive techniques are adopted to record neural activity. MI is the mostly known paradigm for active BCIs. It relies on mental tasks in which the user envisions a movement without actually executing it. Understandably, this is largely studied in rehabilitation protocol, e.g., in presence of a post-stroke brain damage. On the other hand, MI can be used as an alternative way of communication and control.

On these premises, the current chapter introduces the possibility to realize wearable MI-BCI instrumentation. In this case, setups with at least 8–10 electrodes are common, but some studies [371, 372, 373] were conducted to analyze the possibility to reduce the number of acquisition channels and optimize system wearability and portability. Then, neurofeedback modalities are investigated in order to optimize the trade-off between training time and performance, therefore, this concept is introduced hereafter from a metrological point of view.

DOI: 10.1201/9781003263876-9

9.1.1 Motor Imagery

From a metrological point of view, defining the measurand is one of the most challenging issues for an active BCI. Notably, the meaning of "measuring motor imagery" is faced hereafter. Scientific literature indicates that sensorimotor rhythms are brain signals associated with motor activity, e.g., limb movements. They consist of EEG oscillations measurable in the bands μ and β, typically corresponding to the ranges 8 Hz to 13 Hz and 13 Hz to 30 Hz, respectively. The spatial location of SMRs is near the central brain sulcus.

Hence, the information contained in μ and β rhythms is suitable for detecting phenomena associated with motion. Interestingly, they are not merely generated in correspondence of an action, but they also subsist when the action is imagined [374]. This phenomenon implies that people with lost motor function might use MI as a surrogate for physical practice [375, 376]. Indeed, the aim of a MI-based BCI is to measure and classify brain signals related to movement imagination without actually requiring to perform a movement. Typical mental tasks are the imagery of left or right hand movement, but also imaging the movement of feet or tongue. In this sense, MI is an endogenous paradigm for EEG-based BCIs, i.e., it relies on spontaneous and voluntarily modulated brain activity.

Application examples involve motor rehabilitation or the driving of robotic prostheses, but MI-BCIs are also addressed to able-bodied people for communication and control tasks in everyday life. In such cases, it is assumed that the user is able to voluntarily modulate the SMRs. However, as anticipated above, this is rarely the case and proper training is required before being capable of performing with MI. Training protocol (i.e., letting the user practice with the imagination of movements and hence the associated modulation of SMRs) are thus essential in "creating the measurand". Among the different possibilities, neurofeedback can be used as discussed in the following.

9.1.2 Neurofeedback in Motor Imagery

A key challenge in MI is letting the subject "properly imagine the movement", namely generating brain patterns reflecting the imagined movement. In the MI-BCI, the subject can be considered as a transducer between the thoughts (the imagination of a specific movement) and electrical events (the generation of a specific brain pattern). However, this transduction is associated with a high degree of uncertainty, since it is not always certain that the subject will generate the brain patterns associated with that specific movement in a detectable way. In order to make the phenomenon repeatable, reproducible, and hence "metrologically valid", the neurofeedback is a possibility consisting of providing a sensory stimulus in response to the subject's brain activity, thus letting them understanding whether the desired movement has been imagined correctly or not.

Neurofeedback consists of measuring brain activity and giving back to the user an activity-related information. Such an information is usually presented with a visual, auditory, or haptic feedback [377]. In implementing a BCI, the feedback can be part of the system functionality or it can be only exploited during the training process [378, 379]. Studies have shown that self-regulating brain activity through neurofeedback can be beneficial for the BCI performance [380, 381]. Notably, large event-related desynchronization can be observed during the online feedback sessions [382]. Indeed, neurofeedback training determines performance in the successive sessions, but it is still not clear the influence of factors such as given instructions, importance of the mental strategy, aptitude of the user in performing the task, attention to the feedback, and how the feedback is provided.

Many BCI technologies rely on vision for their functionality, and, in particular for the MI-BCI, visual feedback has been employed to enhance the engagement in movement imagination [383]. The advent of VR and AR technologies has given great contribution to that, thanks to the immersive experience offered by a computer-generated scenario [384, 385]. Indeed, it is possible to create virtual objects to interact with, in order to enhance the user engagement in employing an MI-BCI, and hence receive feedback based on the measurement and interpretation of the SMRs. More recently, auditory feedback has been proposed as an alternative for people with deteriorated vision [386] or in conjunction with visual feedback in trying to improve the performance of MI training [387]. These studies investigated different audio settings, from mono to stereo, and even 3-D vector base amplitude panning. They demonstrated that auditory feedback can be a valid equivalent of the visual one. The need to investigate further feedback paradigms arises in aiming to create a more immersive experience, leading to a stronger engagement of the user. In particular, haptic feedback is of great interest for the AR (and VR) community because it enhances the simulation of a real situation [388, 389]. The sensation given by haptic feedback allows the user to better interact with virtual objects by feeling like touching a real object. Starting from that, in a BCI experimental protocol, the haptic feedback can help in user training for MI. This could even be an alternative for people with both visual and auditory impairments, or it could be combined for a multi-modal feedback.

Visual and auditory feedback is already compared [386] showing that there is no statistical difference in their performance. Works regarding vibrotactile feedback are also present in literature [390, 391, 392, 393], though little evidence is present regarding the exploitation of wearable and portable actuators [394]. Therefore, it is worth investigating this feedback modality as well as a feedback combination to enhance the BCI user engagement.

In the following, investigating neurofeedback is discussed in the frame of a wearable BCI instrumentation relying on MI. Notably, visual and vibrotactile feedback is taken into account. The combination of visual and vibrotactile feedback is considered as well. The neurofeedback could to improve MI either during training associated with rehabilitation protocols or for the usage of the instrumentation itself. In the case of users with disabilities, the kind of feedback to provide can be chosen according to the functional possibilities. For instance, vibrotactile feedback could be the only solution for deaf-blind people. In general, the optimal combination of feedback should be used to achieve an enhanced engagement by multi-modal feedback and hence enhance the measurement of MI-related phenomena.

9.2 DETECTION OF EVENT-RELATED (DE)SYNCHRONIZATION

The cortex, the basal ganglia, and the lateral portion of the cerebellar hemisphere are generally involved in planning the movement, and the electrical activity in this region changes according to the motor intention. Without entering into neurological details, the operating principle of an MI-based BCI is to measure both the frequency content and the spatial distribution of the brain activity. Indeed, the different parts of the body are mapped to the primary motor cortex. There is, however, a disproportion among the different body parts in terms of brain mapping, and this gives hints on the most suitable movements to imagine.

For measurement purposes, it is also worth mentioning that each side of the body is controlled by the contra-lateral hemisphere of the brain. Generally speaking, left hand, right hand, and feet movements should be relatively simple to discriminate in recorded brain signals, because the body parts entail a large area of the sensory motor cortex and are spatially separated. Notably, right hand MI produce relevant changes in brain activity in the sensory motor cortex around the electrode position *C3* (see standard locations in Fig. 1.5), while the left hand is associated with the area around electrode *C4*. Foot movement, instead, produces signals on the medial surface of the cortex, around electrode location *Cz*.

Figure 9.1 depict EEG patterns in time-frequency domain as they were measured during left hand, right hand, or feet motor imagery. These spectrograms were calculated by taking into account data from the BCI competition IV (dataset 2a, subject A03T, where T stands for "training session"). In particular, the activity at the standard locations *C3, Cz*, and *C4* are highlighted. As mentioned above, patterns for these imagined movements were expected to be spatially distinct.

The patterns shown in the Fig. 9.1 are examples of *Event-Related Desynchronization (ERD)* and *Event-Related Synchronization (ERS)* , two phenomena, pointed out as responses of neuronal structures in the brain during movement execution or imagination by the neurophysiological literature [53, 396]. These phenomena are typically evaluated in MI-BCI studies to better understand the informative content of the EEG signals under analysis. For instance,

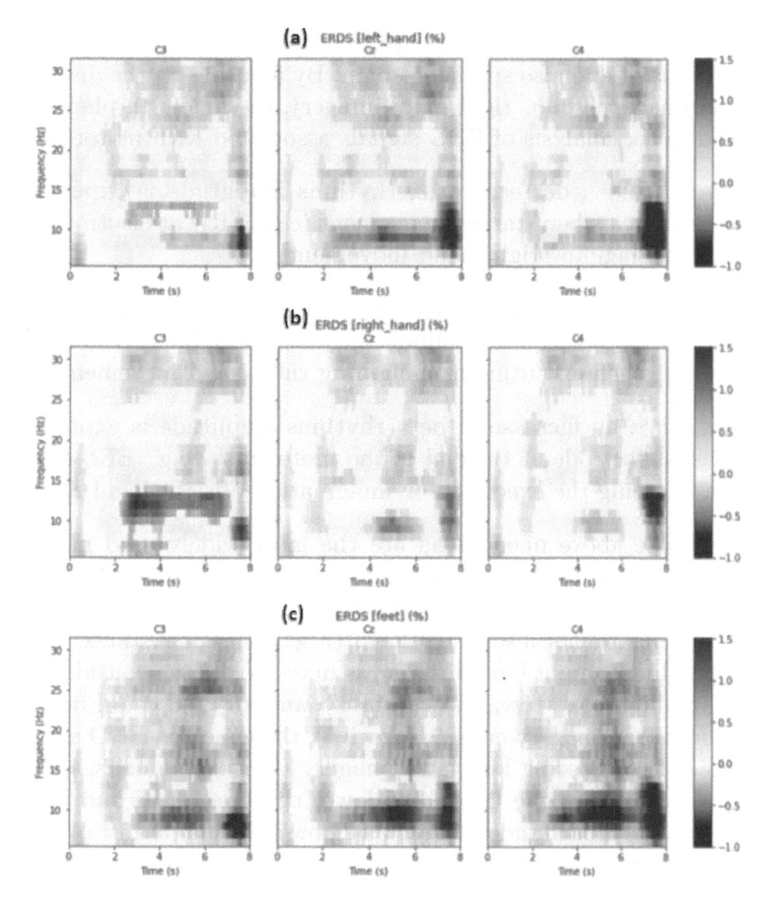

FIGURE 9.1 Time-frequency representation of EEG patterns associated with (a) left hand, (b) right hand, or (c) feet motor imagery, and their spatial distribution, calculated in accordance with [395] with data from subject A03T of BCI competition IV dataset 2a.

ERD/ERS can be analyzed while discussing classification results. Therefore, the next subsection (9.2.1) introduces the neurophysiological processes underlying tasks before thoroughly addressing the calculation of ERD/ERS in Subsection 9.2.2.

9.2.1 Neurophysiology of Motor Imagery

Both sensory stimuli and motor activity induce changes in the activity of neuronal populations [53]. ERD and ERS reflect changes in the activity of local interactions between principal neurons and inter-neurons due to voluntarily modulated spontaneous activity, i.e., without any external stimulus. These phenomena are associated with MI and are investigated in the time domain, frequency domain, and also spatial domain. By focusing on the discrimination of left hand imagery versus right hand imagery, the following phenomena can be detected in the analysis of EEG signals associated with motor execution:

- *alpha-band ERD*: a decrease in α rhythms amplitude is expected at the contralateral area when starting movement (e.g., ERD at electrode *C3* when executing or imagining right hand movement);
- *alpha-band ERS*: an increase in α rhythms amplitude could be expected at the ipsilateral area, almost simultaneously to alpha ERD (e.g., ERS at electrode *C4* when executing or imagining right hand movement);
- *beta-band ERS*: an increase in beta rhythms amplitude is expected at the contralateral electrode at the end of the motor task (e.g., ERS at electrode *C3* when stopping the execution or imagination of right hand movement).

Although the above phenomena are the most widely used in MI, others EEG features are still being studied, such as the contralateral β band ERD emerging when the cue is presented or with the onset of imagination. Moreover, these phenomena are not always detected (Fig. 9.1). For instance, an ERD at *C3* corresponding to right hand imagery is detected for the considered subject data (red area in second row, first columns), and an ERS could be present at *C4* after the motor imagery ends. Meanwhile, the expected ERD and ERS are not detected if considering left hand imagery (first row). Thus, the study of ERD/ERS occurring in the upper and lower rhythms alpha and in the band beta allows us to understand whether and how subjects perform the MI task. Averaging over multiple trials is required to measure such phenomena, and a series of assumptions are necessary: (i) the measurand activity is time-locked with respect to the event, and (ii) the ongoing EEG activity is uncorrelated additive noise. Literature reports that some events may be time-locked but not phase-locked, thus averaging can either highlight a decreased or increased synchronization of neuronal populations [53]. In the following, how to retrieve the time course of event-related phenomena is explained in details.

9.2.2 Time Course of Event-Related Patterns

Several algorithms exist for computing the time course of ERD/ERS. The "power method", originally proposed by Pfurtscheller and Aranibar [53] in 1979, consists of four steps:

1. band-pass filtering of all event-related trials;
2. squaring of the amplitude samples to obtain power samples;
3. averaging of power samples across all trials;
4. averaging over time samples to smooth the data and reduce the variability.

Results of steps 1–3 of this method applied to data from subject A03T are shown in Fig. 9.2. Notably, the right hand motor imagery is taken into account and signals were filtered in the alpha (8 Hz to 11 Hz) and beta bands (26 Hz to 30 Hz) in accordance with [53].

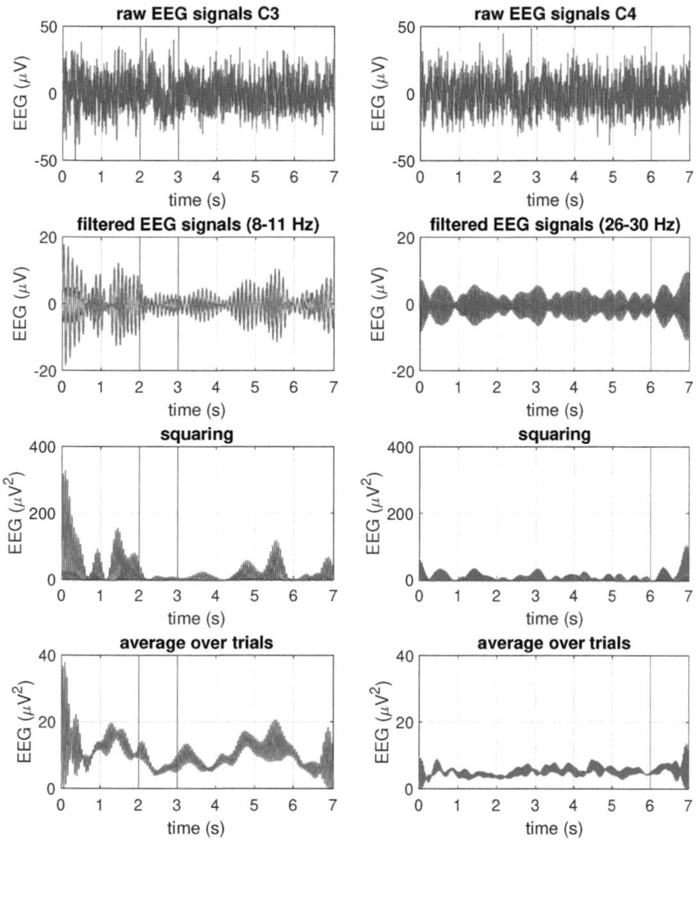

FIGURE 9.2 Calculation of ERD (a) and ERS (b) according to the power method proposed in [53]. A decrease of band power indicates ERD while an increase of band power ERS.

As a disadvantage of the power method, the ERD/ERS could be hidden by event-related activity, which is phase-locked to an external sensory stimulation. Therefore, the "inter-trial variance method" can be also used to

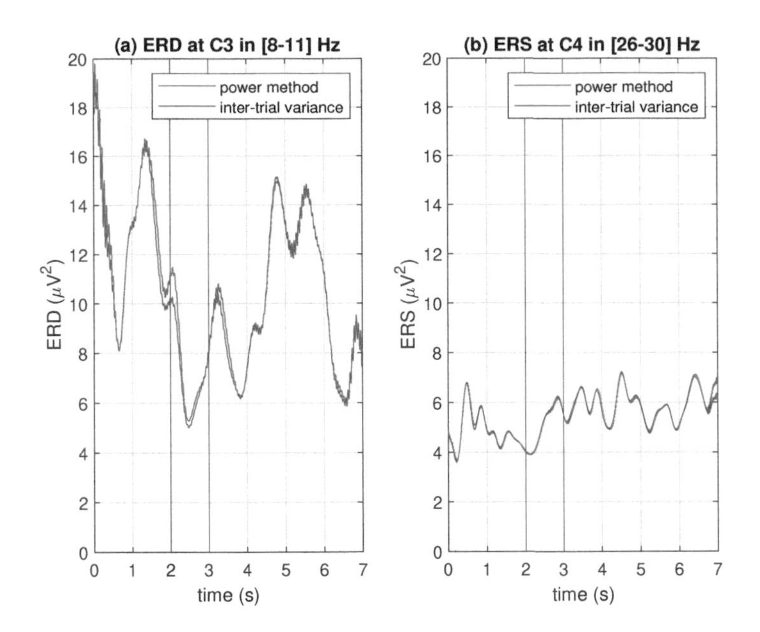

FIGURE 9.3 Comparison of the power method and the inter-trial variance method for calculating the ERD (a) and ERS (b).

distinguish between ERD/ERS and event-related potentials. This consists of three steps:

1. band-pass filtering;
2. calculation of the point-to-point inter-trial variance;
3. averaging over time;

Figure 9.3 compare the ERD/ERS calculated according to the power method with the results of the inter-trial variance method. In the present case, the two methods have only slightly different results. More details on the difference between the two methods can be found in [397].

In the example of Fig. 9.3, the time instants 2 s and 3 s are highlighted as they correspond to the start of the cue indication and of the motor imagery, respectively. A power decrease (ERD) is indeed present in the range from 8 Hz to 11 Hz, while the power increase (ERS) in the range from 26 Hz to 30 Hz is not so clearly identified. Actually, a key aspect of studying ERD is to identify, for each subject, the best frequency bands to calculate the ERD, while, in the previous examples, the chosen bands were selected in accordance with Pfurtscheller and Da Silva [53]. The main methods used for determining the pass band of the filters are:

1. comparison of short-time power spectra: the significant frequency components are calculated by comparing two power spectra of a

reference-trial (R), relative to a reference period, and an event-related trial (A);

2. *Continuous Wavelet Transform (CWT)*: this method involves comparing the signal under analysis with a wavelet ψ dilated and shifted according to a scale parameter; in doing that, a time-frequency representation of the signal is obtained; however, despite a spectrogram relying on the Fourier transform, the CWT allows to analyze high-frequencies with high-time resolution and low-frequency resolution, and low-frequencies with low-time resolution and high-frequency resolution; this is particularly suitable for EEG signals, and it can be used to identify the subject-specific bands associated with lowest (ERD) and highest (ERS) amplitudes;
3. determination of the peak frequency: the ERD is calculated within four frequency bands with a width of 2 Hz, namely theta $[f(i) - 6, f(i) - 4]$, lower 1 alpha $[f(i) - 4, f(i) - 2]$, lower 2 alpha $[f(i) - 2, f(i)]$, and upper alpha $[f(i), f(i) + 2]$; the frequency $f(i)$ depends on the subject and it is identified with an iterative procedure by starting from an initial guess; this aims to avoid to choose the fixed [4, 12] Hz (theta + alpha) band a-priori to customize it on the specific individual.

Figure 9.4 report the ERD/ERS of the previous examples re-calculated with a subject-dependent band selection. According to a time-frequency analysis, the ERS should be better highlighted in the [12, 14] Hz band, but it is not clear in time domain. Instead, the ERD calculated by considering the [11, 14] Hz is much more evident.

Despite the previous examples, the ERD and ERS are typically reported in percentage. This involves calculating the averaged power in a reference period T_{ref} as

$$P_{ref} = \frac{1}{|T_{ref}|} \sum P_t, \tag{9.1}$$

where P_t is the instantaneous signal power, and the ERD is computed as:

$$ERD[t] = \frac{E[P_t] - E[P_{ref}]}{E[P_{ref}]} = \frac{E[P_t]}{E[P_{ref}]} - 1 \tag{9.2}$$

where $E[\cdot]$ is the expected value obtained by averaging across trials. Note that, from equation (9.2), a negative ERD implies that and ERS is present. Indeed, this method assumes a pre-fixed time interval as reference. However, using a dynamic baseline as reference may be preferred. In this regard, a generalized ERD/ERS quantification method was also proposed in 2009 by Lemm, Múller and Curio [398]. This method is more robust in quantifying event-related changes in presence of non-stationary dynamics. The generalized ERD is defined as:

$$gERD[t] = \frac{E[P_t|C = 1]}{E[P_t|C = 0]} - 1 \tag{9.3}$$

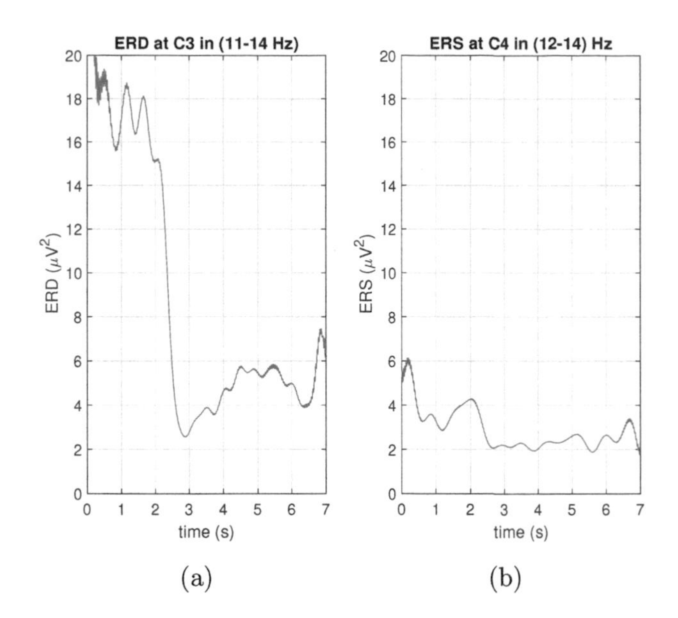

FIGURE 9.4 ERD (a) and ERS (b) calculated according to the power method after choosing subject-specific frequencies. The ERS is still not visible, but the ERD is much more evident.

where $E[P_t|C = 1]$ is the conditional expectation of the band power at time, $E[P_t|C = 0]$ is the unperturbed condition, and C is a binary variable. If $C = 0$ the considered trial relates to an event-free period, Meanwhile $C = 1$ relates to event-related trials. As in the classical case, an ERD corresponds to a decrease in power while an ERS to a power increase. Therefore, although (9.3) is written in terms of ERD, it applies to ERS as well.

Broadly speaking, this method also allows to study the influence of arbitrary factors on the ERD. Further details can be found in [398].

10

Motor Imagery-Based Instrumentation

The present chapter deals with the realization of a MI-BCI system. In doing that, general concepts from current literature are recalled. In details, Section 10.1 presents the design of the system, Section 10.2 reports a prototypical form, and Section 10.3 analyzes the performance of such a system.

10.1 DESIGN

The block diagram of a BCI relying on motor imagery is shown in Fig. 10.1. This basically consists of EEG acquisition and processing. Therefore, no sensory stimulation is needed, in contrast with reactive BCI paradigms. However, neurofeedback could be delivered by relying on online processing of the acquired EEG signals. In such a case, the sensory feedback could be modulated in accordance with the classification of the signals.

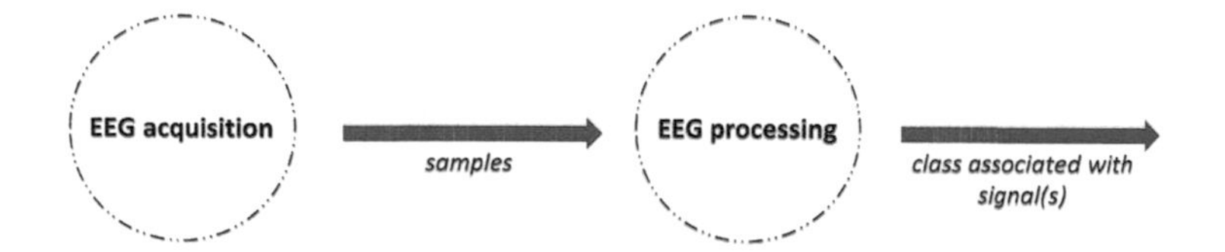

FIGURE 10.1 Block diagram of a simple BCI system based on motor imagery. Neurofeedback could be eventually provided according to the signals classification, but this is not yet represented in this diagram.

In building an MI-BCI for daily usage, the number of electrodes should be minimized, so as to achieve user comfort, wearability, and portability [399]. Possible approaches for choosing electrodes location include data-driven channel selection [400] or knowledge-driven selection [80]. The former typically leads to optimal performance because channel positions can be tuned on the specific subject. However, it requires long experimental sessions in order to acquire enough data. In the latter approach, instead, the user is involved in shorter and less stressful experimental sessions, but final performance is non-optimal since channels are selected according to average

DOI: 10.1201/9781003263876-10

neurophysiological phenomena. Nonetheless, findings from a recent study on channel selection [401] show that data-driven selection is compatible with knowledge-driven one when subject-independent locations are foreseen. Such findings are also discussed in more details in the following chapter. Thereby, data-driven channel selection could be avoided in building a subject-independent system.

A crucial aspect in developing for the MI-BCI is the signal processing approach. A feature extraction step is necessary to process EEG data. In this field, a very successful method is the *Filter Bank Common Spatial Pattern* (FBCSP) proposed in [73]. This approach was successfully replicated and tested on benchmark datasets [73, 402, 401], and some studies even showed its efficacy in processing bipolar channel data [403]. A FBCSP-based classification pipeline is shown in the following. Its performance are extensively discussed in the last part of the present chapter. This constitutes a basis for the ensuing discussion on case studies.

10.2 PROTOTYPE

In realizing a prototype for a wearable MI-BCI, the acquisition hardware is selected first. A preliminary feasibility study demonstrates that the dry electrodes from Olimex [61] have a poor contact impedance when placed at *C3* and *C4* because they are highly sensitive to head movements. The ab medica Helmate [62], instead, has to be discarded because the online data stream is not yet available, though offline analysis demonstrated that the EEG signals quality is good enough even with dry electrodes. Therefore, EEG acquisition is carried out with the FlexEEG headset by NeuroCONCISE Ltd [65] with wet electrodes already introduced in Chapter 2. The headset is used with the FlexMI electrodes configuration, consisting of three bipolar channels over the sensorimotor area. Recording channels have electrodes placed at *FC3-CP3*, *FCz-CPz*, and *FC4-CP4*, according to the international 10-20 EEG standard [30], while the ground electrode is positioned at *AFz*. Electrodes' position and the EEG cap are shown in Fig. 10.2.

Conductive gel is used to ensure low contact impedance and high stability at the scalp interface, though these electrodes could be also used without any gel. The EEG signals from the electrodes are filtered and amplified by the electronic board. Then, these signals are digitized by sampling at 125 Sa/s with 16-bit resolution. Finally, the data are transmitted via Bluetooth 2.0.

The acquired and digitized EEG signals are processed with the FBCSP approach to translate the acquired brain activity into control commands. As represented in Fig. 10.3, the processing pipeline can be divided into four blocks:

1. a *Filter Bank*, extracting the frequency content of EEG signals in different bands;

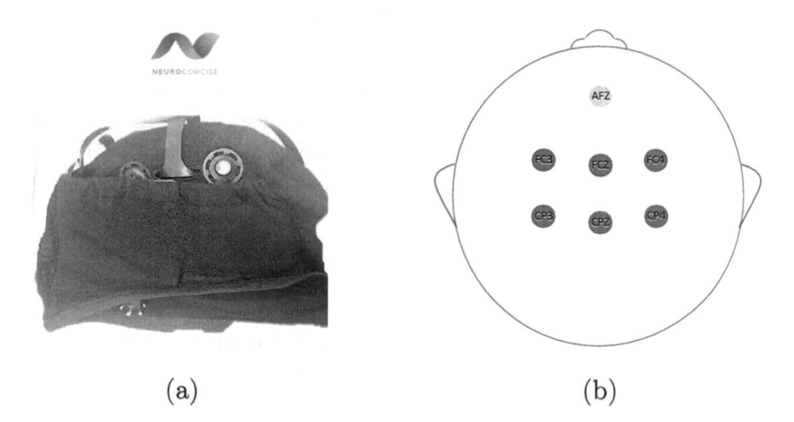

FIGURE 10.2 Wearable and portable EEG acquisition system employed in prototyping a MI-BCI (a) and its electrode positioning (b).

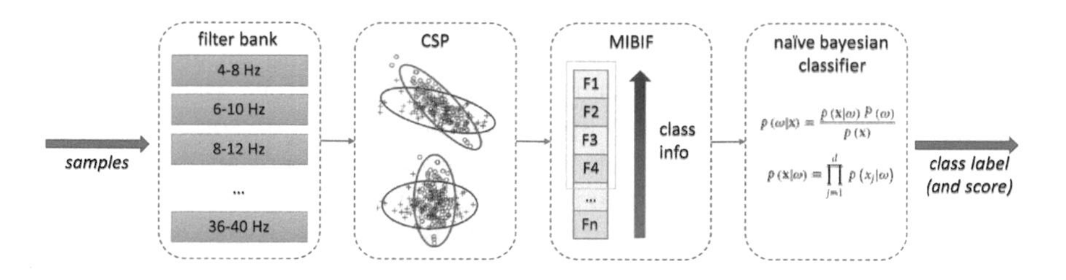

FIGURE 10.3 Filter Bank Common Spatial Pattern for processing EEG signals associated with motor imagery.

2. the *Common Spatial Pattern* algorithm, a widely used features extractor considering EEG spatial information;
3. a *Mutual Information-based Best Individual feature selector* of features accounting for class-related information;
4. an optimal classifier.

Each block is better discussed in the following by describing the particular implementation of the FBCSP processing, along with the training and evaluation phases needed for identifying and testing the algorithm model.

10.2.1 Filter Bank

The first step of FBCSP (already introduced in Chapter 7 and recalled here for the reader's convenience) is the Filter Bank, which extracts the frequency content of each EEG signal with multiple pass-band filters. Chebyshev filters are employed, each one with 4 Hz-wide band and with 2 Hz overlap between consecutive filters. The overall band considered in this implementation are

4 Hz to 40 Hz, thus resulting in 17 bands (4 Hz to 8 Hz, 6 Hz to 10 Hz, 8 Hz to 12 Hz, ..., 36 Hz to 40 Hz).

Despite the original proposal [73], overlapping bands are chosen to avoid losing information in correspondence of band edges, as it may happen in the non-overlapping case. For instance, in the original FBCSP implementation, 8 Hz was the high cut-off frequency and the low cut-off frequency for two adjacent bands, 4 Hz to 8 Hz and 8 Hz to 12 Hz, respectively. To achieve a −200 dB/decade slope of the filter frequency response, the stop-band attenuation and the order are set to −40 dB and to 10, respectively. The high slope is needed to have a sharp cut between pass band and stop band, but this also implies a significant computational burden for filter application to many EEG data. The filter bank is applied to each EEG signal, therefore, different channels are treated separately. At the end of this step, the number of signals is 17 times the number of the original signals.

10.2.2 Spatial Filtering

The *Common Spatial Pattern* (CSP) extracts features relying on the covariance matrix of measured EEG signals. The spatial content associated with EEG signals is projected in a new space, so that features are sorted according to the class-related variance [404]. For instance, if data are available for left and right hand motor imagery, two matrices for spatial projection can be computed: the first leads to features with left-related variance in descending order, while the second leads to features with right-related variance in descending order. Note that, in this binary case, maximizing the variance associated with a class automatically minimizes the variance associated with the other class. Therefore, the two matrices are linked. After the spatial projection, only most relevant features are taken into account. In this sense, the CSP is a spatial filter.

The CSP computes projection matrices from available data, therefore, two phases can be distinguished: the matrices are calculated from data during the training phase, while they are applied to data in the evaluation phase. The projection matrix computation is as follows. Given a $n_{ch} \times n_{samp}$ matrix E_j of EEG signals per each experimental trial, where n_{ch} is the number of channels while n_{samp} is the number of samples, the covariance matrix associated to the class c is calculated as:

$$K_c = \frac{1}{n_{tr,c}} \sum_{j=1}^{n_{tr,c}} \frac{E_{j,c}E'_{j,c}}{trace\left(E_{j,c}E'_{j,c}\right)}, \tag{10.1}$$

where $n_{tr,c}$ is the number of trials available for class c and the apostrophe ($'$) indicates the transposition of a matrix. The matrix K_c is actually obtained as a mean of the covariance matrix associated with the trials $j = 1, 2, ..., n_{tr,c}$, each of which is normalized by the respective matrix trace. Summing the K_c

matrices of the classes, the composite covariance matrix

$$K = \sum_{c} K_c \tag{10.2}$$

is then obtained. Note that $c = 1, 2$ in the simple binary case, but Eq. (10.1) and Eq. (10.2) still hold in the multi-class case.

A complete projection matrix W_c is computed per each class by solving the eigenvalue decomposition problem:

$$K_c W_c = K W_c \Lambda_c, \tag{10.3}$$

where eigenvalues are the non-zero values of the diagonal matrix Λ_c, while W_c is made of eigenvectors. If eigenvalues are sorted in descending order, the eigenvectors are sorted accordingly. In the final step, the actual projection matrix W_c^r is retrieved from W_c by considering only the first and last m columns. The former has maximum variance with respect to class c, while the latter have minimum variance with respect to class c. Ultimately, the CSP training phase consists of calculating the W_c^r for all the classes of interest.

In the evaluation phase, instead, the CSP projection is applied. The matrices W_c^r transform the $n_{ch} \times n_{samp}$ data associated to each trial following the equation

$$C_{j,c} = W_c{}' E_j E_j{}' W_c, \tag{10.4}$$

and then features are obtained as:

$$\mathbf{f}_{j,c} = \log \left[\frac{diag(C_{j,c})}{trace(C_{j,c})} \right]. \tag{10.5}$$

where the $C_{j,c}$ matrix is diagonalized, and $2m$ features are obtained per trial. Spatial content of EEG data is filtered because $2m < n_{ch}$. Hence, the array $\mathbf{f}_{j,c}$ synthetically describes the $j - th$ trial with respect to the class c. If the CSP is exploited after the filter bank, this reasoning has to be repeated for each band. By merging the features of all bands, each trial is described with $2mf$ features (with regard to class c).

The literature demonstrate the effectiveness of CSP in extracting discriminatory information from two populations of motor imagery data [405], but multi-class extensions are feasible as well [404]. With the CSP trained and applied according to Eqs. (10.1)–(10.5), a possible approach is the One-Versus-Rest (OVR). In the latter approach, the binary discrimination is extended by considering each class against the remaining ones.

10.2.3 Features Selection

Other than feature extraction, the CSP naturally encompasses feature selection. However, multiple bands are derived from the filter bank, therefore, a further selection step is needed in choosing only the best features combining spatial and frequency content.

The selection approach considered is the *Mutual Information-based Best Individual Feature* (MIBIF) [406], which relies on class-related information associated with each feature. This reduces the $2mf$ features representing each trial to $n_{M,c}$ features. The subset of selected features depends on the class c. Also in this algorithm step, data-driven training is needed before the actual selection. Notably, the mutual information between a class and a feature is calculated, in the binary case, as:

$$I_c(f_i) = H_c - H_{c|f_i} = -\sum_{c=1}^{2} P(c)\log_2 P(c) - \sum_{c=1}^{2}\sum_{j=1}^{n_{tr}} p(c|f_{j,i})\log_2 p(c|f_{j,i}), \tag{10.6}$$

which is the information related to a feature $i = 1, 2, ..., 2mf$ with respect to class c is obtained by subtracting from the class entropy H_c the feature-related entropy $H_{c|f_i}$. The latter is derived with the conditional probability between a feature and a class summed over trials, estimated with the Bayes rule

$$p(c|f_{j,i}) = \frac{p(f_{j,i}|c)P(c)}{\sum_{c=1}^{2} p(f_{j,i}|c)P(c)}. \tag{10.7}$$

The a-priori probability $P(c)$ of a class is estimated with the frequentist approach as the ratio between the $n_{tr,c}$ trials available for the class and the total number of available trials n_{tr}. The conditional probability $p(f_{j,i}|c)$ is estimated with the Parzen Window [73]:

$$\hat{p}(f_{j,i}|c) = \frac{1}{n_{tr,c}}\sum_{k=1}^{n_{tr,c}} \frac{1}{\sqrt{2\pi}} e^{-\frac{\left(f_{j,i}-f_{k,i}\right)^2}{2h^2}} \tag{10.8}$$

Remarkably, in the sum, a feature $f_{j,i}$ associated with the $j-th$ trial appears in the difference with each features $f_{k,i}$ from the same class, and it is weighted by the smoothing parameter

$$h = \left(\frac{4}{3n_{tr}}\right)^{1/5}\sigma, \tag{10.9}$$

with σ equal to the standard deviation of $f_{j,i} - f_{k,i}$.

By these calculations, the I_c values associated with each of the $2mf$ features are obtained. They are then sorted in descending order so as to select the first $k_{MIBIF} = 5$ ones, i.e., the most informative ones. The number of features is empirically chosen after some preliminary trials and then fixed for next elaborations. Actually, the CSP features are paired (first m ones versus last m ones), therefore, when one feature is selected, one must also select the paired one. For this reason, the effectively selected features $n_{M,c}$ can range from 6 to 10. Overall, the MIBIF chooses the best features according to the data-driven training, while in the evaluation phase these same features are selected. Therefore, the band choice is subject-related if a subject-by-subject training is carried out. Finally, the MIBIF algorithm here presented for two classes can be extended to more classes, for example with a pair-wise approach (discrimination among pairs of classes) or with the already mentioned OVR approach.

10.2.4 Classification of Mental Tasks

In the last algorithm step, the features are classified. A Supervised Approach is exploited, hence training is needed before the evaluation phase. Again, the classifiers are introduced for a binary case but they can be extended to more classes. The first FBCSP implementation [73] proposed a *Naive Bayesian Parzen Window* (*NBPW*) classifier, but also the SVM is investigated. Broadly speaking, the classifiers lead to compatible performance. While the SVM is extensively described (Section 7.2.2), it is useful to discuss the NBPW implementation.

The idea behind the Bayesian classifier consists of calculating, for a trial, the probability of a class given the features $\bar{\mathbf{f}}$ describing that trial. The Bayes rule

$$p(c|\bar{\mathbf{f}}) = \frac{p(\bar{\mathbf{f}}|c)P(c)}{\sum_{c=1}^{2} p(\bar{\mathbf{f}}|c)P(c)} \tag{10.10}$$

is applied. While the $P(c)$ is obtained with the frequentist approach on training data, the conditional probability is computed with a "naive assumption", according to which all features are conditionally independent, so that

$$p(\bar{\mathbf{f}}|c) = \prod_{i=1}^{n_{M,c}} p(\bar{f}_i|c). \tag{10.11}$$

Each of these conditional probabilities are estimated with the Parzen window of (10.8) and (10.9) already introduced for the MIBIF selector. The training of the NBPW consists of using features associated with training trials in the Parzen window expression, as well as in the frequentist estimation of $P(c)$. After the probability $p(c|\bar{\mathbf{f}})$ is calculated per each class, the most probable class is assigned to the trial during the evaluation phase.

The FBCSP approach with NBPW or SVM classifiers is implemented in Matlab and tested on benchmark datasets from BCI competitions. Such datasets were created in different European institutes to let BCI researchers compete in developing the best processing approach. After the competitions [17, 407, 72] (the last one was in 2008), these data were published and they have been largely used in research activity. Thanks to that, different research groups can compare their results, though the lack of standardization in this community still prevents their full interpretation and replication.

The benchmark dataset used in this book is presented hereafter, and inherent results are then presented. However, few expert subjects were usually involved in acquiring these data, therefore, the real usability of a motor imagery BCI in daily life could not be simply evaluated on this dataset, and statistical significance of eventual improvements could not be proven due to the small subjects sample. For these reasons, in the following sections more data will be considered to better evaluate the proposed FBCSP variants.

10.3 PERFORMANCE

In the previous part of this chapter, great attention was put on the processing of EEG signals associated with motor imagery. This complies with the fact that the processing is a crucial part of the measurement of brain activity, especially when dealing with motor imagery detection.

According to the previous discussion, an offline analysis is conducted to assess the performance of the FBCSP algorithm. Such an analysis is reported hereafter by firstly recalling benchmark dataset, widely exploited in the literature. Using largely spread data for testing an algorithm is indeed a well-known practice in the Machine Learning community, and this is also in accordance with the need to make experiments reproducible, which is inherent in metrology and in the science as a whole. Hence, classification accuracy is investigated as well as the presence of ERDS in the brain signals under analysis, in accordance with the discussion of Chapter 9. This will be then also useful for online analysis and, therefore, for neurofeedback applications.

10.3.1 Benchmark Datasets

The first dataset used for testing the FBCSP implementation is the *dataset 2a from BCI competition IV* (2008) [72]. It includes EEG data from nine subjects recorded through 22 channels in two different days (sessions). Data were sampled at 250 Sa/s. Each session was composed of six runs separated by short breaks. A run consisted of 48 trials balanced between motor imagery classes, i.e., “left hand”, “right hand”, “feet”, and “tongue”. Hence, 12 trails per class were available. In processing these data, cross-validation was applied to data from the first session (identified as A0xT for subject “x”) in order to predict the classification accuracy expected on further data. In a second step, these data were used for training the pipeline, while the second session (A0xE) was used for the evaluation.

The second dataset used in this book is the *dataset 3a from BCI competition III (2006)* [407], including 60-channels data from only 3 subjects. Data were sampled at 250 Sa/s. Per each class 60 trials were available, and the classes were “left hand”, “right hand”, “single foot”, and “tongue”.

Two more datasets from BCI competitions are then considered, namely the *dataset 2b of BCI competition IV* [408] and the *dataset 3b of BCI competition III*. These include nine and three subjects, respectively, but the peculiarity is that data were recorded from three bipolar EEG channels and neurofeedback was provided during the trials. Again, data were sampled at 250 Sa/s. Details about the number of available trials are not recalled here, but they will be recalled later when relevant. However, it is to remark that these data were split in half for training and test, as already explained for the first dataset.

For all the four datasets, a cue-based paradigm was adopted. This meant that in each trial the user is relaxed until an indication appears on the screen. After about 1 s from the indication, the user had to carry out the specific mental task until a stop indication was provided. In this sense, this was a synchronous BCI paradigm. The feedback, when present, was provided on the same screen.

10.3.2 Testing the Feature Extraction Algorithm

The first evaluation of the FBCSP performance is done on dataset 2a with a six-folds cross-validation on training data. The six possible class pairs are considered as well as the four classes altogether.

Results are reported in Tab. 10.1 for the NBPW classifier case. In particular, the mean classification accuracy among subjects is reported along with the associated standard deviation to also have an estimate of the mean dispersion. Notably, the results are compatible with the ones obtained by an SVM with linear kernel, which are not reported in the table. In the binary cases, the accuracy goes from 74 % to 84 % depending on the considered pair, while in the four classes case it is 63 %. Directly comparing results achieved for different numbers of classes is unfair because the more the classes and the easier the misclassification. Hence, other metrics exist for normalizing the classification accuracy to the number of classes (e.g., see the Cohen's kappa [409]). Despite that, the classification accuracy will be still considered in the following as it is the most diffused one.

TABLE 10.1 Mean and standard deviation of the cross-validation accuracy obtained on dataset 2a with a Bayesian classifier.

Tasks	**ACC ± STD (%)**
left hand vs right hand	74 ± 20
left hand vs feet	81 ± 13
left hand vs tongue	82 ± 13
right hand vs feet	81 ± 15
right hand vs tongue	84 ± 13
feet vs tongue	75 ± 13
four classes	63 ± 19

The relatively high values for the standard deviations indicate that there is much performance difference among subjects. Figure 10.4(a) shows box-plots related to the cross-validation accuracies of the nine subjects from dataset 2a with the addition of the three subjects from dataset 3a.

These results are shown for the same classes set as before, and again only the NBPW-related results are shown. In addition, Fig. 10.4(b) reports classification accuracies obtained by training on the first half on data and evaluating on the other half (i.e., the hold-out method).

In both the figures, the dashed line represents the random accuracy level, i.e., the theoretical accuracy that would be obtained if one guesses at random the classes. By assuming balanced classes, this level equals 50 % in the two-classes cases and 25 % in the four-classes case.

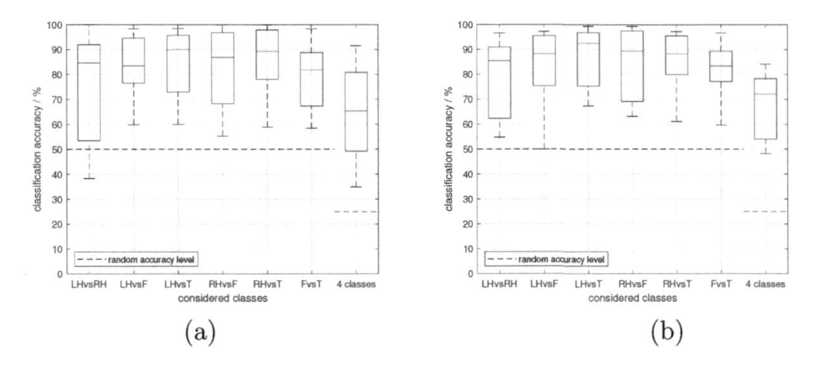

FIGURE 10.4 Classification accuracies for the 12 subjects from dataset 2a and dataset 3a. The six possible classes pairs are considered as well as the four classes case: cross-validation on training data (a) and hold-out method (b).

The above results show how the median accuracies (red lines) are significantly higher than the random accuracy level, and that only in few cases the accuracy distribution goes below that level. Overall, classification accuracy is relatively high if compared to other datasets or more real applications, as it will be seen later in this book.

Results are also reported for the 12 subjects from dataset 2b and dataset 3b. Only two classes are available, thus Tab. 10.2 reports the accuracy of all subjects by comparing cross-validation results with hold-out ones. Only two bipolar channels out of three were considered in these tests. Despite the low channel number, the results are compatible with the previous ones, namely the performance is unexpectedly high. Indeed, later results will show that reducing the number of channels is typically associated with accuracy decrease. On the one hand, the results of Tab. 10.2 may point out that datasets from BCI competitions are not always representative for real application scenarios. On the other hand, these data were acquired while providing neurofeedback during MI.

TABLE 10.2 Classification accuracies for the 12 subjects from dataset 2b and dataset3b. Subject-by-subject accuracies are reported for cross-validation (CV) and hold-out (HO).

	Accuracy (%)	
Subjects	**CV**	**HO**
B01	72	67
B02	61	59
B03	54	57
B04	92	96
B05	84	88
B06	80	82
B07	72	74
B08	78	89
B09	80	85
S4b	80	79
X11b	75	74
O3VR	65	64
mean	74 ± 10	76 ± 12

10.3.3 ERDS Detection

After the classification results, it is interesting to assess the presence of ERDS in the brain signals under analysis and associate it with classification accuracy.

In assessing the ERDS, the python MNE (MEG + EEG analysis and visualization) package [410] is used.

An ERDS map is obtained, namely, a time/frequency representation of the phenomenon [411]. Epochs are extracted from the initial relax period to the break after motor imagery, time-frequency decomposition is applied to all epochs by means of a Multi-Taper Transform, then a cluster test is conducted so that only epochs with p-value less than 5 % would be considered for averaging, and finally time-frequency maps are obtained per each electrode by averaging over considered epochs.

As a second step, the time course of ERDS could be derived from the time-frequency map by considering alpha and beta bands. In this book, this is done for channels *C3* and *C4*, on the left and right side of the somatosensory area, respectively.

In addition to the ERDS, also the time course of classification accuracy can be investigated. The FBCSP implemented in Matlab is taken into account for the classification of MI-related signals. In doing that, trials are segmented according to the timing of the experiment. The classification should be random in the relax periods and non-random in the motor imagery periods. Hence, if trials are analyzed with a sliding window, one should expect a random level for the classification accuracy at the beginning, an increase in the motor imagery period, and again a random classification in the final break.

In addition to that, in the adopted FBCSP implementation, selected features could be analyzed. Cross-validation is considered, and histograms are built with the features selected across all iterations.

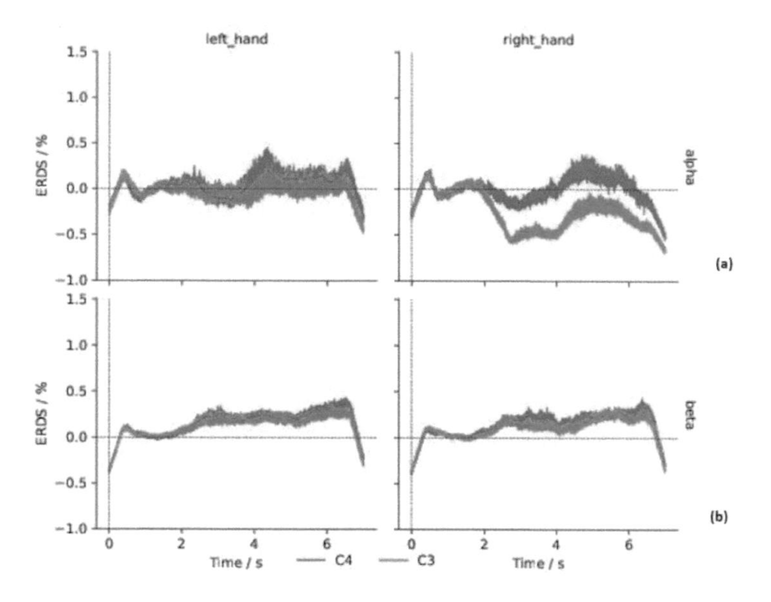

FIGURE 10.5 Time course of ERDS in alpha (a) and beta (b) bands for subject A03 (session T) from BCI competition IV dataset 2a.

These analyses are first made on a benchmark dataset, the BCI competition IV dataset 2a. Analyzing EEG data associated with subject A03, the ERDS time course, in the alpha and beta bands, of Fig. 10.5 is obtained. An ERD in the alpha band is clearly visible for the right hand imagery at the contralateral electrode (*C3*). This ERD is localized at about $t = 2\,\mathrm{s}$, corresponding to the cue indicating the task to perform. Then, an ERS seems present before $t = 6\,\mathrm{s}$, where the imagination should be stopped, though this is still in alpha band (not beta band). No other phenomenon among the abovementioned one appears evident. Moreover, these phenomena are not even clear for the left hand imagery.

Next, the time course of classification accuracy for the subject A03 (Fig. 10.6(a) shows that the discrimination between left and right hand imagery is random in correspondence of the starting relax period and then it increases up to almost 100 % during the motor imagery period. Hence, A03 can be considered a "good" subject in the sense of classification. In the final part, the accuracy also decreases as it should (ending relax period).

In addition to that, the analysis of features is shown in Figure 10.6(b). The histogram highlights that, across the several iterations, the selected features are mostly concentrated in the alpha band. This is in accordance with the

ERD phenomenon appearing in the alpha band for the present subject and it also points out a regularity in the features selected for classification.

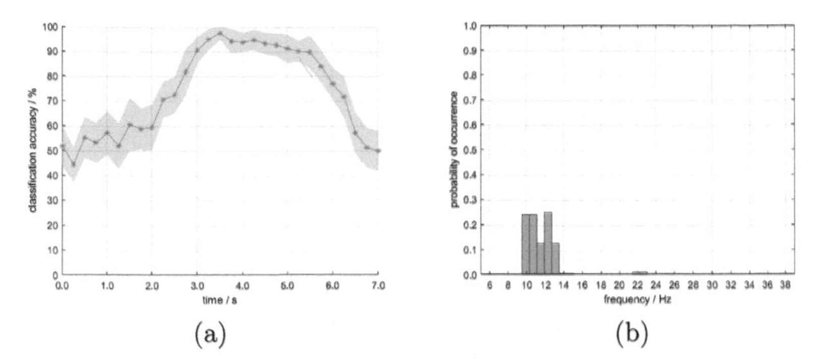

FIGURE 10.6 Analysis of MI-related phenomena for subject A03 (session T) from BCI competition IV dataset 2a: time course of classification accuracy (a) and mostly selected features (b).

The same analyses can be repeated for the subject A05. Fig. 10.7 show that no clear ERD nor ERS phenomenon is visible. However, the time course of classification accuracy (Fig. 10.8(a) points out that a classification accuracy up to 90 % is obtained during the motor imagery period. Hence, also A05 is associated with a good classification performance.

Finally, the histogram of selected features reported in Fig. 10.8(b) still shows concentrated features in the beta band, though they are more diffused than the previous case.

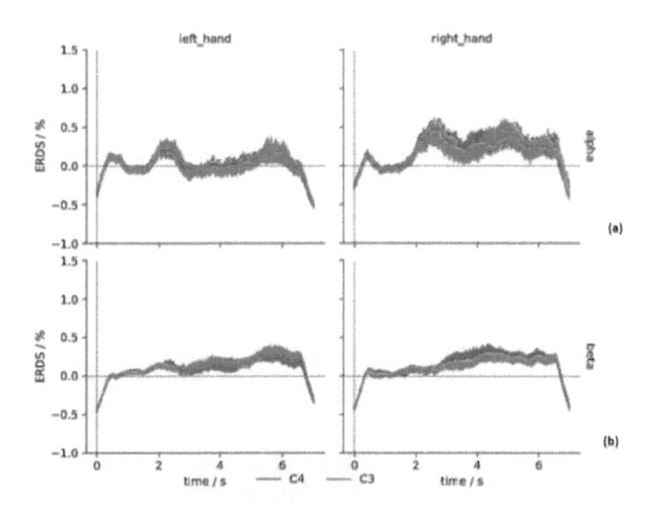

FIGURE 10.7 Time course of ERDS in alpha (a) and beta bands (b) for subject A05 (session T) from BCI competition IV dataset 2a.

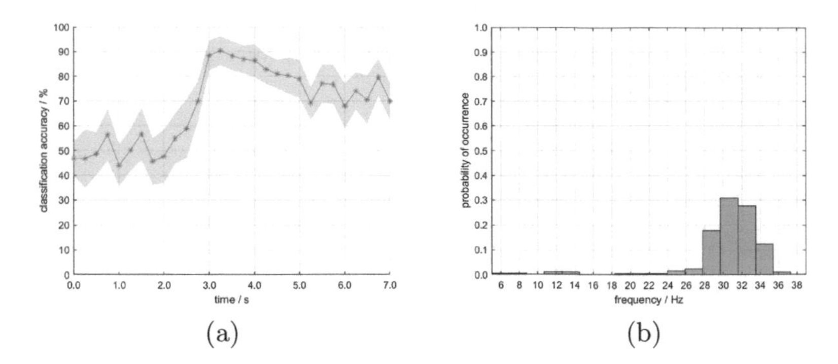

FIGURE 10.8 Analysis of MI-related phenomena for subject A05 (session T) from BCI competition IV dataset 2a:time course of classification accuracy (a) and mostly selected features (b).

As a last case, the ERDS associated with subject A02 is reported in Fig. 10.9. Here, an anomalous behavior appears in the alpha band for both left and right hand imagery. ERS seems present at the cue time instant, while ERD could be identified about in correspondence of the motor imagery stop.

The analysis of classification accuracy in time reveals that there is random classification over the whole period (Fig. 10.10(a), and even the histogram does not show any regularity, but selected features are spread over the allowed frequencies (Fig. 10.10(b).

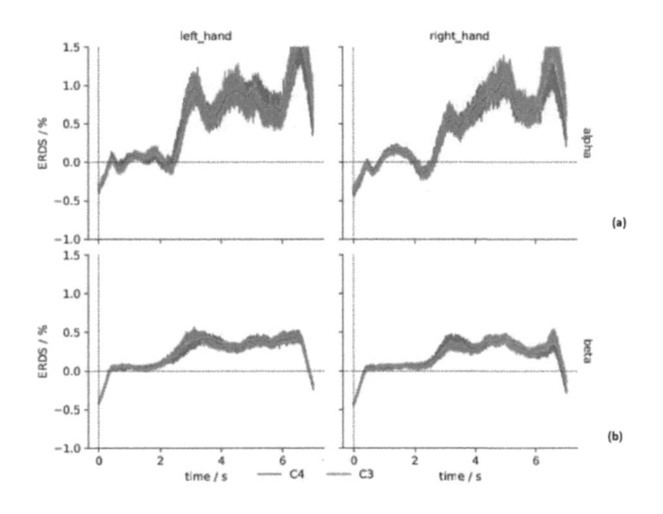

FIGURE 10.9 Time course of ERDS in alpha (a) and beta (b) for subject A02 (session T) from BCI competition IV dataset 2a.

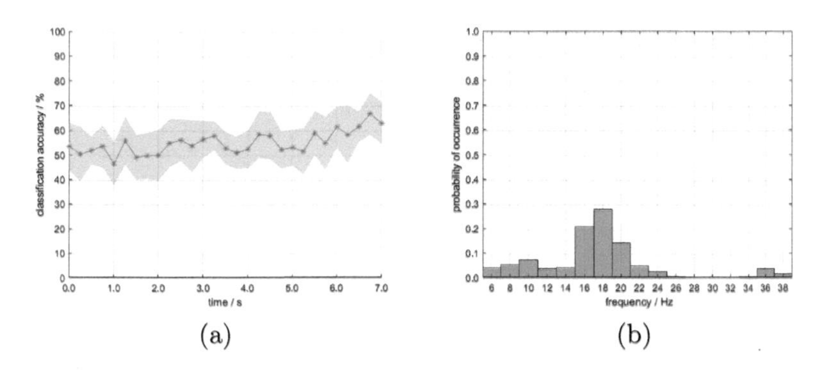

FIGURE 10.10 Analysis of MI-related phenomena for subject A02 (session T) from BCI competition IV dataset 2a: (a) time course of classification accuracy, and (b) mostly selected features.

11

Case Studies

In this chapter, motor imagery-based instrumentation is discussed with reference to twofold case studies. In Section 11.1, the possibility to optimize wearability in a system for control applications is treated.

In Section 11.2, user-machine co-adaptation is presented by exploiting neurofeedback for training the user in imagining movements, while the processing algorithm is trained for recognizing the user's motor intention from EEG. Such a system could be used in neuromotor rehabilitation.

11.1 WEARABLE SYSTEM FOR CONTROL APPLICATIONS

In satisfying the requirements of daily-life Brain-Computer Interfaces, wearability and portability are strictly linked to the number of channels employed in measuring the EEG. Selecting a minimum number of channels not only enhances system wearability and portability, but also optimizes the performance by reducing over-fitting and excluding noisy channels [412]. Differently from the SSVEP case, many channels are usually needed to map the spatial information associated with motor imagery. Moreover, it is not trivial to choose electrodes locations a-priori. For these reasons, the number and location of the acquisition channels are firstly investigated with a data-driven approach. In doing that, the FBCSP algorithm introduced above is mainly exploited. Then, also a knowledge-based approach is investigated, with the aim to choose a small number of subject independent channels.

11.1.1 Experiments

An attempt with a single channel

In pursuing utmost wearability and portability for the MI-BCI, the possibility to exploit a single-channel EEG is firstly attempted.

This investigation is inspired by a research of 2014 [82] proposing the classification of four motor imagery tasks with single-channel EEG data. In their work, the authors proposed the usage of a short-time Fourier transform to

DOI: 10.1201/9781003263876-11

obtain spectrograms from single-channel EEG signals, and then the application of the CSP for features extraction. The CSP would require multi-channel data, therefore, the idea behind that work was to exploit the different frequency bands of the spectrogram in spite of channels. A single channel was selected in post-processing among the available ones, and 3 s-long time windows were processed in order to classify the MI tasks.

In replicating the proposed approach, the *short-time Fourier transform* is calculated in Matlab with the "spectrogram" function, by considering 100 samples-long windows with 50 % overlap between consecutive windows. Each one is zero-padded to 128 samples and a Hamming window is used. Considering the module of the spectrogram, a real matrix is associated to each single-channel signal. From the spectrogram, if frequency bands are considered in place of channels, the CSP can be applied for extracting features. Finally, these features are classified with an SVM. Note that data-driven training is needed for both the CSP and SVM.

The authors of [82] reported a mean classification accuracy equal to 65 % when classifying four motor imagery tasks from the dataset 3a of BCI competition III. The maximum accuracy, obtained in a very specific condition, was about 88 %, and 4 s of the input EEG signals were analyzed.

One attempt of the present book is thus to replicate such results even on a further benchmark dataset, i.e., the dataset 2a of BCI competition IV. The analyses are conducted first by fixing the number of CSP components to $m = 2$ in accordance with the FBCSP approach introduced earlier, and then by also varying the number of CSP components, as the authors of [82] proposed. Classification results suggest that in classifying two motor imagery tasks with data from the dataset 3a, the maximum accuracy that can be obtained is about 85 % (subject k6b, right hand vs tongue). However, all the accuracies associated with subjects from dataset 2a are compatible with the random accuracy, i.e., 50 %. Figure 11.1 shows the results with respect to the nine subjects of the dataset 2a when considering two motor imagery tasks to classify. The random accuracy level is also reported as a red dashed line which is practically superimposed to the lines indicating classification results. Then, if considering four classes, accuracy is even lower. This evidence, together with the fact that no other relevant literature work attempts the usage of a single channel for MI, discouraged further investigation in this direction. Instead, such results triggered the need to study the trade-off between the number of channels and the achievable classification accuracy, as it is reported in the following subsection.

EEG channels selection

In selecting channels with a data-driven approach, the contribution of each of them to motor imagery classification has to be estimated. Therefore, a method is proposed for selecting and then validating the selection of EEG channels. The method exploited the FBCSP classification approach [73] and, to evaluate

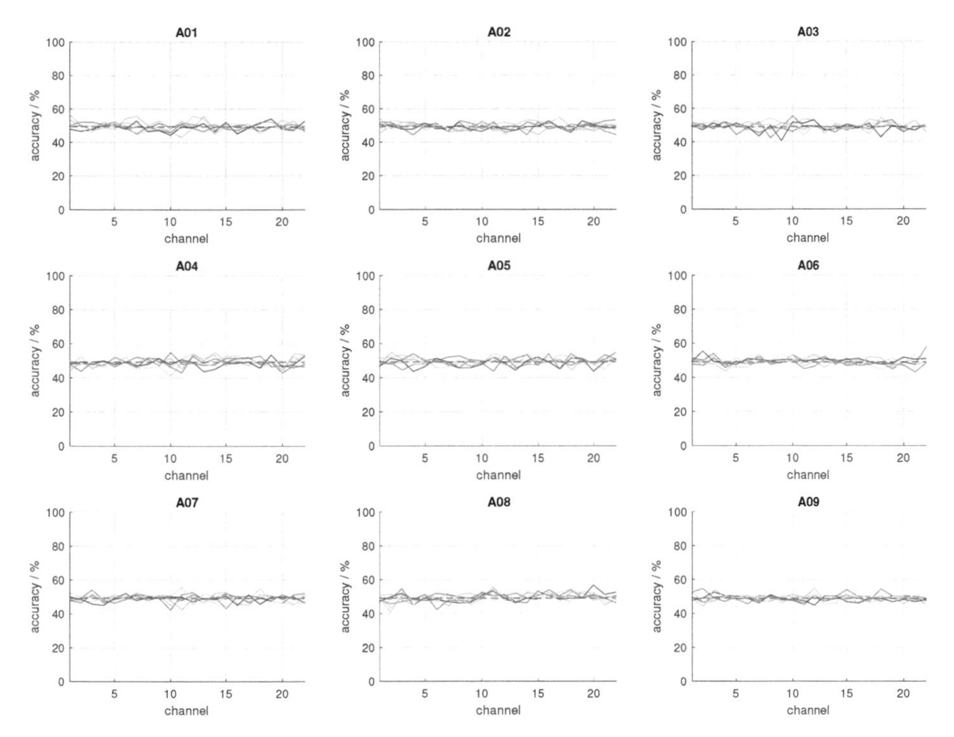

FIGURE 11.1 Results of motor imagery classification with single channel data from dataset 2a of BCI competition IV. The six possible pairs of classes are considered for the nine subjects: left hand vs right hand (blue), left hand vs feet (red), left hand vs tongue (yellow), right hand vs feet (magenta), right hand vs tongue (green), and feet vs tongue (cyan).

the contribution to the final performance per each channel, a Non-Uniform Embedding strategy [413] is added.

According to [412], which distinguishes between different selection approaches, the proposed method is a wrapper technique, because, in contrast with filtering approaches, classification is involved in the selection process.

Overall, the method consists of a progressive channel selection, thus allowing to retrieve the trade-off between the number of channels and the classification performance.

In more details, the selection step involves an iterative process known as *Sequential Forward Selection Strategy* [33], so to choose the best-performing channels: firstly, motor imagery classification is attempted with every single channel in order to select the best one, and then the other available channels are added one-by-one again according to classification performance. Therefore, n channels are used in the n-th iteration: the first $n-1$ are the best from previous iterations, while the last one is found from the remaining channels by assessing the classification accuracy resulting from an n-channel set. The iterations are stopped when reaching the maximum number of available channels. Moreover, the classification performance at each iteration is assessed as the mean among all available subjects. By doing that, the method attempts to find subject-independent best channels. The difference between mean classification accuracy and associated standard deviation among the subjects ($\mu - \sigma$) is maximized in order decide for the best performance. Note that this kind of objective function tries to also minimize the performance variation (σ). To ease the comprehension of the selection algorithm, this is represented in Fig. 11.2.

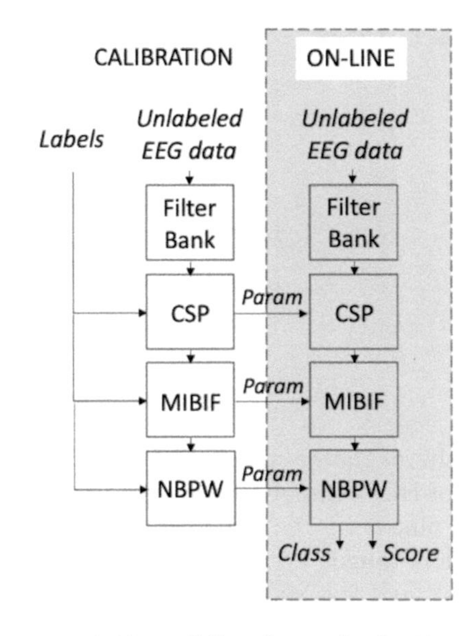

FIGURE 11.2 Representation of the channel selection algorithm exploiting the FBCSP approach for the classification of EEG signals.

The channel selection step corresponds to the training of the system, and clearly it should only exploit training data. Therefore, a six-fold cross-validation is used for the performance assessment. The training data are thus split six times in six-folds, of which five-folds are used for training the FBCSP and the remaining one for calculating the accuracy. Per each subject, cross-validation accuracy is obtained as the mean across six-folds, and then mean and standard deviation across the subjects can be obtained. Thanks to the cross-validation, the method aims to determine the most significant channels in terms of predictive information.

In using the FBCSP approach, a suitable classifier is also found by comparing state-of-the-art solutions, such as the NBPW [73], the SVM, and the kNN classifier [33]. The classifier choice is again data driven. these binary classifiers were extended to multi-class with the "one-versus-res" approach. Hence, each class is discriminated against the remaining ones, and the class assigned to each trial is the one with the highest classification score.

At the end of the channel selection step, a sequence of sorted channels is available. Channels are sorted according to the predictive information evaluated with cross-validation, and these results are useful in designing a BCI system because the expected classification performance (μ, σ) is given as a function of the channels subset. However, the channel sequence should be validated on new (independent) data. Hence, a testing phase is needed to validate the BCI design by also taking into account the trained FBCSP with the classifier. Such a validation i conducted by simply considering possible channel subsets according to the found sequence. In the first test step a single channel is considered, while for the following steps a channel is added each time according to the channels order. Note that, instead, it would be unfair to select channels by relying on test data.

The results for the channel selection and validation are reported in the following.

11.1.2 Discussion

The channel selection method illustrated here is applied to the benchmark datasets of BCI competitions 2a (nine subjects) and 3a (three subjects) introduced earlier.

The method is implemented in MATLAB. First, the optimal classifier is chosen by considering both binary and multi-class (four classes) classification. Table 11.1 reports the comparison results for the NPBW, SVM, and kNN classifiers in terms of μ and σ for the nine subjects of the dataset 2a. The six possible pairs of classes were considered for the binary cases (combination of "left hand", "right hand", "feet", "tongue").

The cross-validation accuracy is calculated on data from the *sessions T*, thus choosing the classifier regardless of evaluation data data (sessions E) to be employed later. All channels (22 for dataset 2a) are taken into account in this step. Performances appear compatible for the different classifiers if looking at the intervals defined by mean and standard deviation. Nonetheless, paired t-tests [414] are performed to have an objective criterion for the choice.

Tests are conducted by considering two classifiers per time for a binary classification problem or for the multi-class case. The null hypothesis for each paired test is that the mean accuracy μ associated with a classifier is equal to the one associated with the other. Therefore, rejecting the null hypothesis would suggest a difference in the performance in terms of mean accuracy. Then, in calculating the t-statistic, also the standard deviation is taken into account. The level of significance α for the test is fixed at 5 %.

In these conditions, there is no strong statistical evidence to prefer a classifier over the others. However, hyperparameters are tuned for the SVM and the kNN, while this is not needed in the NBPW case. This implied a small preference for the NBPW, and in addition the Bayesian approach is preferred

TABLE 11.1 Comparison between NBPW, SVM, and kNN classifiers for different classification problems. Mean cross-validation accuracy and associated standard deviation were calculated among nine subjects (dataset 2a) by taking into account all channels.

	ACC ± STD (%)		
Tasks	NBPW	SVM	kNN
left hand vs right hand	74 ± 20	74 ± 20	73 ± 19
left hand vs feet	81 ± 13	82 ± 12	82 ± 13
left hand vs tongue	82 ± 13	81 ± 13	82 ± 13
right hand vs feet	81 ± 15	81 ± 15	81 ± 15
right hand vs tongue	84 ± 13	83 ± 14	83 ± 13
feet vs tongue	75 ± 13	76 ± 13	75 ± 14
four classes	63 ± 19	63 ± 19	63 ± 20

since it naturally gives back a probability as classification score, which could indicate class uncertainty.

After the classifier selection, the channel selection procedure is carried out. The results are reported in Figs. 11.3 and 11.4, in terms of mean cross-validation accuracy (blue line) and standard deviation of the mean $\sigma_\mu = \sigma/\sqrt{9}$ (blue shaded area). On the x-axis, the channels that are progressively selected within the proposed procedure are reported, while the y-axis reports the classification accuracy.

A direct comparison between Fig. 11.3 and Fig. 11.4 would not possible because the y-axes should report a performance normalized by the number of classes. Such a metric could be the kappa coefficient [409], but classification accuracy is still considered since it is more common in literature. In trying to overcome this issue, the y-axes values reported in the figures are such that there is a one-by-one correspondence between Fig. 11.3 and Fig. 11.4, namely they correspond to the same Cohen's kappa coefficients.

The plots show the trade-off between channels and classification performance. In some cases, an acceptable performance can also be achieved with four channels, especially in the binary classification problems. Meanwhile, more channels are typically required in the four-task case. Hence, the figures highlight the possibility of suitable minimization for the number of channels while preserving the desired classification performance. Focusing on the four classes, for instance, the same accuracy of the 22 channels case can be achieved with 18 channels, but the accuracy corresponding to 10–12 channels could be still acceptable while gaining in wearability and portability. In some other cases, the maximum accuracy seems to be reached with less than 22 channels. Although this could indicate that some noisy channels had been removed, in this case those values were a random occurrence since they did not result (from the statistical point of view) significantly different from the accuracy with the whole number of available channels.

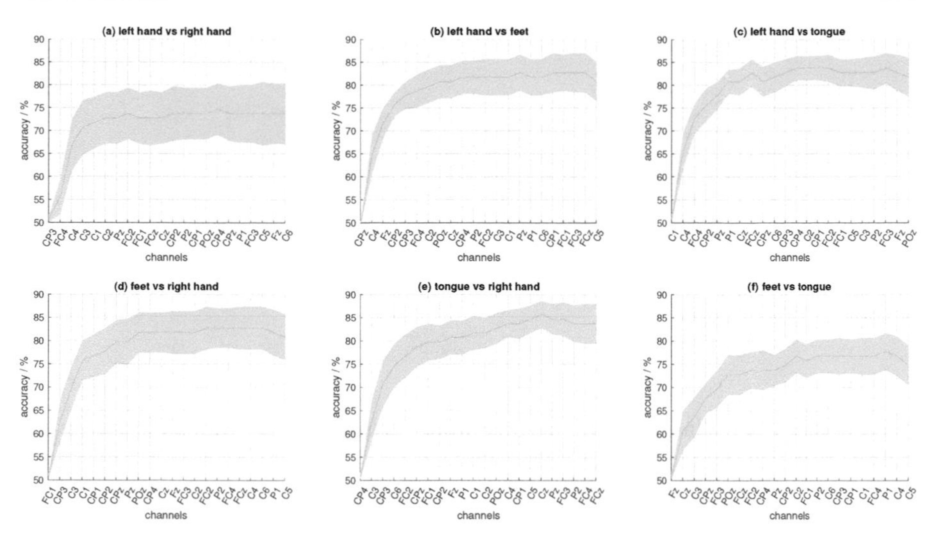

FIGURE 11.3 Classification performance (cross-validation) for progressively selected channels and for each pair of classes: left hand vs right hand (a), left hand vs feet (b), left hand vs tongue (c), right hand vs feet (d), right hand vs tongue (e), and feet vs tongue (f).

The results of the channel selection step are resumed in Tab. 11.2, where the classification performance at the maximum number of channels (22 in this case) is compared to the performance obtainable with a reduced number of channels, which is here chosen as eight channels as a reasonable trade-off between performance and user-friendliness for the final system.

TABLE 11.2 Mean and standard deviation of the classification accuracy obtained during channel selection, for both 8 and 22 channels.

	ACC ± STD (%)	
Tasks	**8 channels**	**22 channels**
left hand vs right hand	74 ± 17	74 ± 20
left hand vs feet	81 ± 11	81 ± 13
left hand vs tongue	83 ± 9	82 ± 13
right hand vs feet	80 ± 15	81 ± 15
right hand vs tongue (12)	80 ± 11	84 ± 13
feet vs tongue (11)	74 ± 12	75 ± 13
four classes (18)	57 ± 18	63 ± 19

The sequences of channels are validated in a further step by employing the independent data from *sessions E*. As anticipated, for each pair of classes and for the multi-class problem, the algorithm is trained on data from the first session before the second session data could be classified.

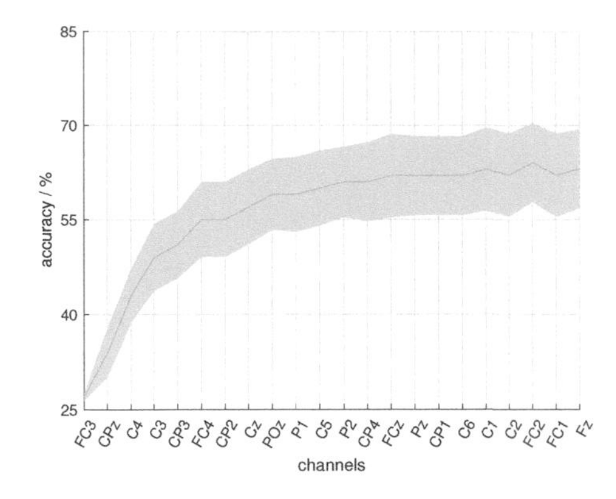

FIGURE 11.4 Classification performance (cross-validation) for progressively selected channels in the four-class problem.

The results are plotted in Fig. 11.5 and Fig. 11.6. The channels on the x-axes correspond to the respective sequence, which are found during the selection, while on the y-axes the accuracy values are chosen as before. As a whole, validation results are compatible with the results obtained during channel selection, but more fluctuations are present with respect to the almost monotonic behavior of the channel selection. Therefore, the results of Fig. 11.5 and Fig. 11.6 are analyzed with paired t-tests. The null hypothesis is that the accuracy at the maximum number of channels equals the accuracy at a reduced number of channels, so rejecting the null hypothesis would mean that the performances are different, either better or worse at the maximum number of channels. The level of significance is again set to $\alpha = 5\%$. Failing to reject would not mean that the null hypothesis is necessarily true. Therefore, when rejection is not possible, the probability β of a false positive is also taken into account, so that $\beta \leq 5\%$ is considered as a reasonable risk of accepting the null hypothesis. These tests highlighted that performances are significantly worse than 22 channels ones when 3 to 5 channels are considered. Meanwhile, performances become acceptable with 6 to 13 channels, depending on the considered classification problems.

Validation results are resumed in Tab. 11.3 by reporting the classification performances for both the reduced number of channels (highlighted in the table) and the maximum number of available channels. In accordance with previous tables, also this table reports the standard deviation σ in spite of the standard deviation of the mean $\sigma/\sqrt{9}$ plotted in the figures.

As a further validation of the proposed method, data from session T and E are flipped in order to repeat the channel selection and channel selection steps. Apart from an accuracy diminishing of about 1 %-5 %, the channel selection proves still effective in selecting a smaller number of channels while accepting a known accuracy diminishing. Hence, flipping data from the two sessions suggest that the results shown above are not restricted to the particularly

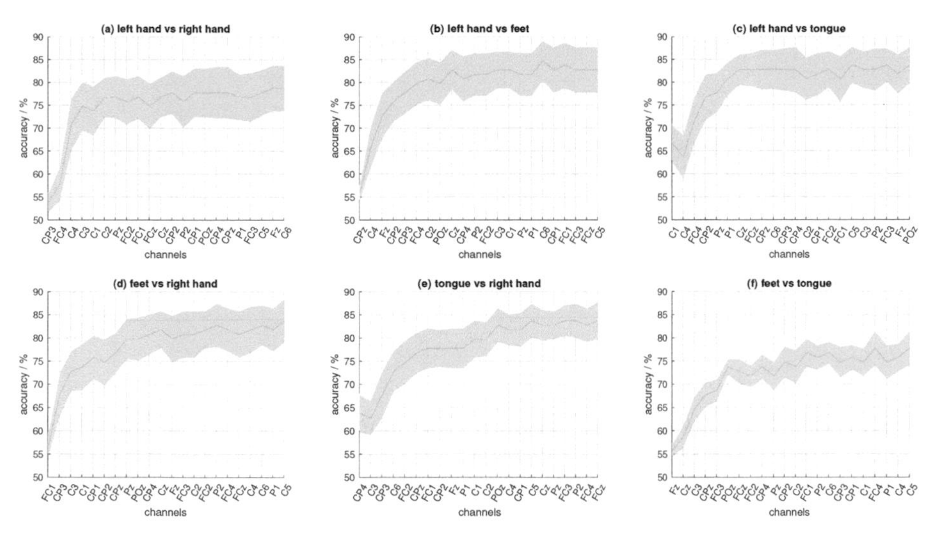

FIGURE 11.5 Mean classification performance obtained to validate the respective sequence of the channel selection procedure in the binary classification cases for the six possible pairs of classes: left hand vs right hand (a), left hand vs feet (b), left hand vs tongue (c), right hand vs feet (d), right hand vs tongue (e), and and feet vs tongue (f).

chosen data. In going further, however, also the dataset 3a is used. With the FBCSP approach plus NBPW, classification accuracy for all the available channels (60) results above 80 % for two tasks, and above 70 % for four tasks, while the accuracies for reduced sets of channels result between 78 % and 92 % for in the binary cases, and about 72 % with 10 channels in the four tasks case.

These results are also plot in Fig. 11.7 by considering a representative example of the binary cases ("right hand vs tongue") and the four classes case.

TABLE 11.3 Mean and standard deviation of the classification accuracy obtained during channel sequences validation.

	ACC ± STD (%)	
Tasks	**Reduced channels**	**22 channels**
left hand vs right hand (6)	77 ± 12	79 ± 15
left hand vs feet (6)	80 ± 15	83 ± 15
left hand vs tongue (6)	81 ± 12	84 ± 12
right hand vs feet (10)	81 ± 12	84 ± 15
right hand vs tongue (13)	83 ± 12	84 ± 12
feet vs tongue (13)	77 ± 9	78 ± 12
four classes (10)	62 ± 12	64 ± 12

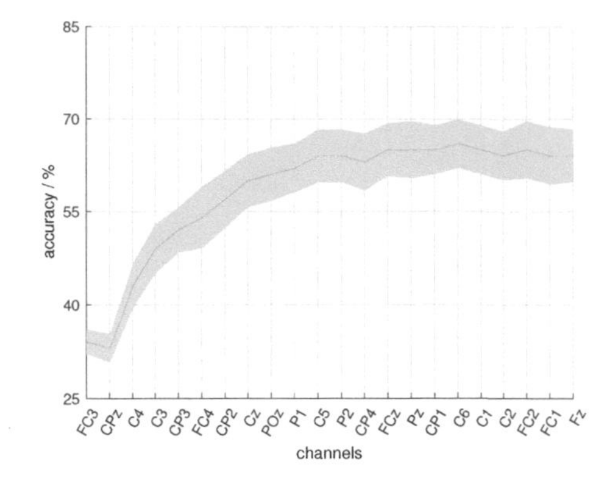

FIGURE 11.6 Mean classification performance obtained to validate the respective sequence of the channel selection procedure in the four-classes case.

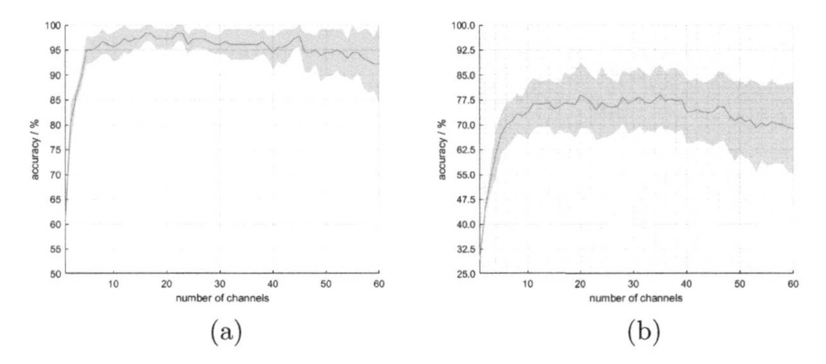

FIGURE 11.7 Classification performance in validating the channel sequences found on dataset 3a: right hand versus tongue (a) and four classes (b).

In general, the results presented here are compatible or better than the findings in the recent literature [415, 416]. However, results can be criticized under some aspects. Firstly, although considering datasets from BCI competitions guarantees reproducibility of the results, classification accuracies in real applications are usually lower because only the best subjects were selected for acquiring these datasets. Moreover, the results also point out the need for further improvements. For instance, the fluctuations of the mean accuracy in validating the channel sequences is probably addressable to the non-stationarity of EEG signals. Indeed, thanks to the modular channel selection involving FBCSP, the algorithm can be refined, especially for managing the non-stationarity. In addition to that, the iterative selection itself could be also enhanced. In particular, figures report an accuracy diminishing for some selection steps, and then an increase if one or more other channels are added. This should suggest that channels are correlated. Therefore, correlation-based selection could improve the results [417].

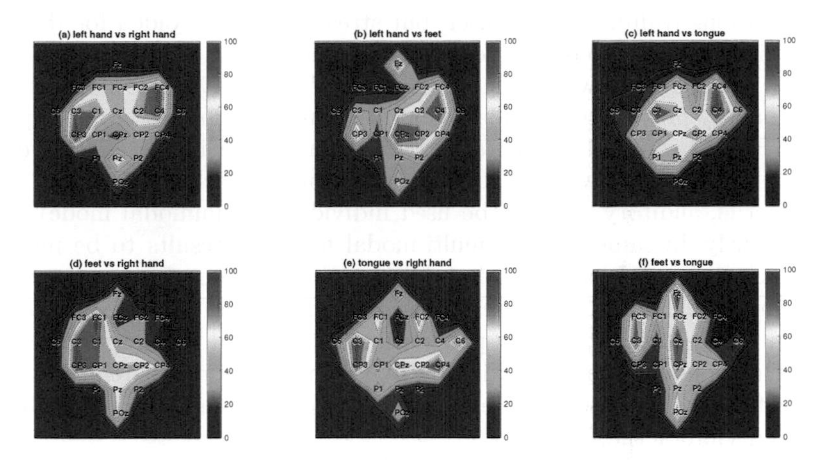

FIGURE 11.8 Most predictive information on the scalp for each pair of classes: left hand vs right hand (a), left hand vs feet (b), left hand vs tongue (c), right hand vs feet (d), right hand vs tongue (e), and feet vs tongue (f).

In conclusion, the location on the scalp of predictive information is analyzed. According to the literature, the right hand is related to contralateral activation, the left hand generally shows a bilateral activation, the tongue is related to the interhemispheric fissure of the sensorimotor cortex, while feet are related to a strong bilateral activation [418, 419]. Hence, Fig. 11.8 show contour plots for the binary classifications in which a weight is assigned to each channel of a sequence: maximum weight is assigned to the firstly selected channel, while the weight progressively diminishes as the channels are selected later. With this reasonable assignment of weights, the areas highlighted by a concentration of most predictive channels are effectively in agreement with the above-mentioned literature.

11.2 USER-MACHINE CO-ADAPTATION IN MOTOR REHABILITATION

As mentioned in the first chapter, in an *active BCI*, the subject voluntarily produces a modulation of the brain waves for controlling an application, independently of external events. In particular, a BCI based on motor imagery is mainly used in control external devices such as a wheelchair, robotic prosthesis, or virtual objects (including virtual hands) in an XR application. The main disadvantage of this application is the need to train the subject to “imagine correctly”. Also, the longer the system operation, the higher the performance should be due to more available data (for algorithm training) and long training for the user, but stress must be avoided for the user, since this would be deleterious for the performance.

In this framework, *Neurofeedback* (*NF*) helps the subject to pursuit this goal, by providing a feedback about his imagination. In control applications, NF can be notably exploited for improving motor imagery classification.

It is well known that different types of feedback can be used, i.e., visual, tactile, auditory and can be used individually (unimodal mode) or simultaneously. In same studies, multi-modal feedback results to be more effective than unimodal one [420]. In addition, the creation of a proper virtual environment allows to increase the users' engagement. This is particularly useful in rehabilitation field: the patient is more engaged and performs better [421]. Therefore, a NF multimodal MI-BCI solution for motor-rehabilitation applications is discussed in the following. The proposed system combines visual and vibrotactile feedback.

The designed and implemented BCI recalls the general architecture already shown in Fig. 10.1 but the FBCSP-based classification algorithm is executed only and the classification output is used for feedback modulation. The designed feedback consisted of controlling both intensity and direction of a moving virtual object. Multimodal feedback is obtained by merging visual and vibrotactile modalities. In detail, the visual feedback is provided by the rolling of a virtual ball on a PC screen, while the vibrotactile one is given by a wearable suit with vibrating motors. Intensity and direction of the feedback were determined by means of the user's brain activity, measured through EEG.

In this prototype, a generic PC monitor is used to provide the visual feedback. However, this will be replaced by smart glasses to provide a more immersive experience and hence furtherly increase user engagement. An Unity application is purposely developed to have a virtual environment with a rolling ball, as well as to control the haptic suit. Note that the applications also indicated the task to carry out (Fig. 11.9). The hardware for the haptic feedback

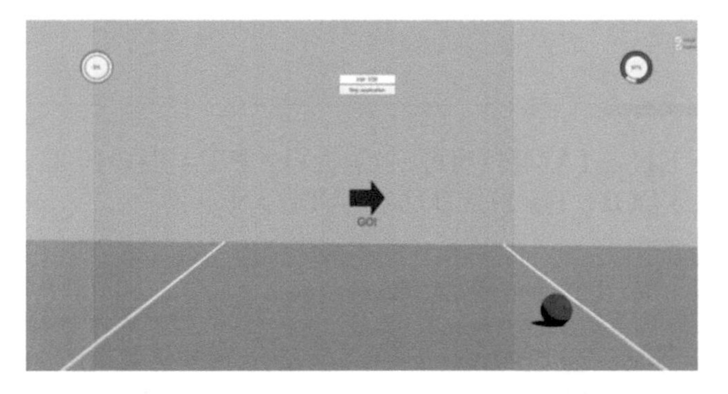

FIGURE 11.9 Visual feedback consisting of a mentally-controlled virtual ball [422].

consists of a vibrotactile suit from bHaptics Inc. This suit is indeed wearable and portable and it is commercially available for gaming [423]. It provides a double 5×4 matrix with vibration motors installed on the front and back of the torso. Vibration can be adjusted in terms of intensity per each single motor, so that patterns can be created to give a specific haptic sensation to the user. The suit can communicate through Bluetooth with a computing unit. In this application, it is controlled from a PC that receives the EEG data through UART and then sends control commands to the suit according the the EEG classification. The suit and the motors locations are shown in Fig. 11.10.

FIGURE 11.10 Wearable and portable haptic suit with 40 vibration motors [422].

This system is validated with some experiments, where three feedback modalities have to be compared: only visual, only vibrotactile, and visual plus vibrotactile (multimodal). Regarding online classification, a training is first executed with data from pure motor imagery trials to identify the classification model. Then, the EEG has to be classified during the motor imagery execution in order to provide a concomitant feedback. Therefore, the FBCSP approach with the Naive Bayesian classifier is applied to a sliding window. The width of the window could range from 1 s to 3 s and it is chosen on training data so as to maximize the classification accuracy. Then, the shift of the window is fixed so as to have a feedback of 250 ms.

The training of the algorithm considers the optimal motor imagery time window. The model performance after training is validated with a cross-validation procedure.

The adopted protocol is a classical synchronous one. The timing of a trial is shown in Fig. 11.11 and it is recalled from the common paradigm of BCI competitions [72]. It consists of an initial relax (2 s), a cue indicating the task to carry on, the motor imagery starting at t =2.5 s, and then its ending at t =6 s. A final break/relax is also present and its duration is randomized between 1 s and 2 s. Though the EEG stream is continuously processed, during the experiments the feedback could be actually provided starting from the cue. Thanks to the Bayesian classifier, a class score could be naturally associated with the class assigned to the processed EEG data because it assigns a probability to each EEG data epoch.

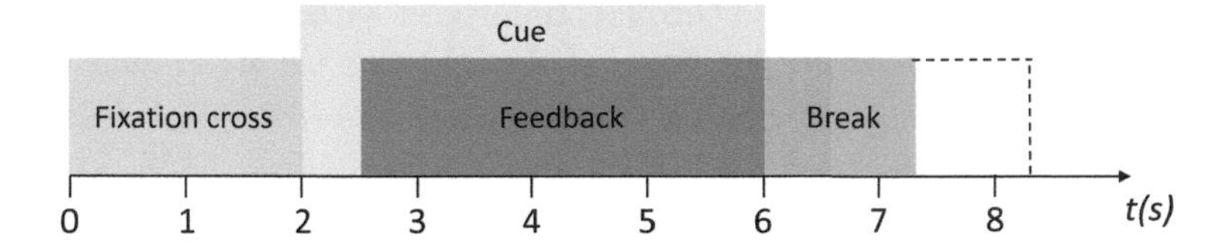

FIGURE 11.11 Timing diagram of a single trial in the BCI experiment with neurofeedback.

11.2.1 Experiments

Different experimental sessions are carried out on different days. Each session is organized in two blocks. In the first block, EEG data are acquired while the user executed motor imagery tasks without any feedback. These data are then used to identify a model for online classification of EEG during the next block. Therefore, in the second block, the user executed motor imagery tasks while receiving feedback about the adherence of his/her EEG activity to the identified model. The EEG data stream is processed with a sliding window approach. The width of the window and the overlap between adjacent windows is also decided during the model identification. The output of online classification consists of a class and a score. In each trial, the feedback is actually provided only if the retrieved class is equal to the assigned task, while the score is used to modulate feedback intensity.

The neurofeedback block is actually divided into three sub-blocks where the three feedback modalities are provided in a randomly different order per each subject. Moreover, left and right motor imagery is also randomly assigned during trial execution to avoid any bias. The described neurofeedback protocol is resumed in Fig. 11.12.

First experiments involve four subjects (three males and one female) and the only multimodal feedback modality. Two subjects perform two series of 40 trials each while the other two perform fewer trials due to technical problems. Subjects are required to wear the suit to receive haptic feedback and the helmet for EEG acquisition. In this preliminary experiment, data are analyzed offline to assess the effectiveness of the multimodal feedback. Specifically, two different procedures for accuracy assessment are carried out. Firstly, a four-fold cross-validation with five repetitions is considered, i.e., the data are divided 5×4 times between training and test set and accuracy is assessed for each group. Then the mean accuracy and standard deviation are considered

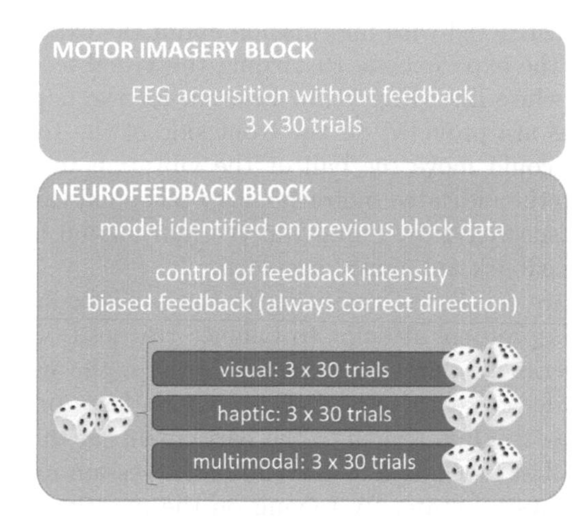

FIGURE 11.12 Blocks and sub-blocks of the neurofeedback protocol employed in the closed-loop motor-imagery-BCI.

in order to assess the performance. The second procedure consisted in evaluating the accuracy by dividing the data in half and assigning 50 % of it to the training set and 50 % to the test set.

Results are shown in Tab. 11.4 and this preliminary analysis demonstrate how the multimodal XR neurofeedback might help the user during the motor imagination. However, experiments with a greater number of subjects are necessary to better validate the system design.

TABLE 11.4 Classification results for the preliminary experiments with four subjects [422].

Subject	Cross-Validation Accuracy	Standard Deviation	Evaluation Accuracy
S1	63 %	14 %	68 %
S2	84 %	6 %	65 %
S3	40 %	19 %	50 %
S4	69 %	8 %	70 %

Therefore, a further experiment with a larger sample size is conducted. Eight right-handed volunteers (SB1-SB8, three males, mean age 26 years) participated in the experiment. No subject reports medical or psychological illness and/or medication and they have normal or corrected to normal vision. Subject SB1 is experienced in active, passive and reactive BCI paradigms, subject SB2 is experienced in active BCI, subjects SB3 and SB4 are experienced in reactive BCI, while the other four subjects have no BCI experience.

Subjects are instructed with information about the experimental protocol before beginning the experiments. Regarding the visual feedback, the goal is to overcome the white line (Fig. 11.9). Instead, for the haptic feedback, the vibration patterns are provided on the front side of the torso starting from the center. They could move the ball or the vibration to the left or to the right according to the indicated motor imagery task, and the goal for the user is to reach the respective side. Finally, in the multimodal feedback case, the aforementioned feedback are jointly provided.

Experiments are conducted according to the neurofeedback protocol proposed above. This protocol attempt to balance the need of much EEG data with limited experiment duration to avoid stress to the user. Two or three sessions are recorded for each subject. Each session last about 2 h.

A questionnaire is administered to the participants at the beginning of the session, after the first block, and at the end of the session in order to monitor the mental and physical state. By relying on the questionnaire proposed by [424], the questionnaire of Tab. 11.5 is proposed in the current experiments.

At the beginning of the first session, the participants are also instructed on the movement imagination itself. They are asked to try different ways of imagining hands movement (such as kinesthetic sensation, squeezing a ball, grasping an object, snapping their fingers, imagining themselves or another person performing the movement) to identify the one they are most confident with. Once chosen, they are asked to keep it constant throughout the session. Finally, they are instructed to avoid muscle and eye movements during the motor imagery task.

As already mentioned, the first block consisted of acquiring EEG data with no feedback provided in order to identify the classification model. Then, three runs with 30 trials each and two classes of imagery are recorded while providing feedback. Participants are asked to imagine the movement of the right or left hand. The order of the cue-based tasks is randomized to avoid any bias. A maximum of 10-min break is given between the two blocks. In the meanwhile, the selection of the best time window to train the algorithm for the online experiment is carried out. For this purpose, the FBCSP is exploited in a ten-repeated 5-folds Cross Validation technique. Its performance is tested from 0 s to 7 s with an overlap of 0.25 s. A time window 2-s wide is used to extract the EEG signal. The time-varying classification accuracy and associated standard deviation is calculated from the time-varying classification accuracies obtained using the cross-validation setup. For each subject, the best time window is chosen in terms of maximum classification accuracy during the motor imagery task together with the minimum difference between accuracies per class. Therefore, the algorithm is trained by using such an optimal window. Notably, another run could be recorded if the results are not satisfactory. Finally, between feedback sub-blocks, a break of about 2 min is given. With 30 trials per run, the total number of available trails per each subject is 360 acquired under different conditions. In analyzing data offline, the scope is to highlight the effectiveness of a feedback modality in improving the

TABLE 11.5 Questionnaire provided to the participants at each experimental session.

Experimental Information St start	
Date	yyyy:mm:dd
Session	
Starting time	hh:mm
Handedness	1: left / 2: right / 3: both
Age	
Sex	1: male / 2: female
Do you practice any sport?	0: no / 1: yes / 2: professional
BCI experience	0: no / 1: active / 2: passive / 3: reactive / 4: multiple types
Biofeedback experience	0: no / number: how many times
How long did you sleep?	number: hours
Did you drink coffee within the past 24 h?	0: no / number: hours before
Did you drink alcohol within the past 24 h?	0: no / number: hours before
Did you smoke within the past 24 h?	0: no / number: hours before
How do you feel?	Anxious 1 2 3 4 5 Relaxed
	Bored 1 2 3 4 5 Excited
	(Physical state) Tired 1 2 3 4 5 Very good
	(Mental state) Tired 1 2 3 4 5 Very good
Which motor imagery are you confindent with?	1: grasp / 2: squeeze / 3: kinesthetic
After training block	
How do you feel?	(Attention level) Low 1 2 3 4 5 High
	(Physical state) Tired 1 2 3 4 5 Very good
	(Mental state) Tired 1 2 3 4 5 Very good
Have you nodded off/slept a while?	No 1 2 3 4 5 Yes
How easy was motor imagery?	Hard 1 2 3 4 5 Easy
How do you feel?	(Attention level) Low 1 2 3 4 5 High
	(Physical state) Tired 1 2 3 4 5 Very good
	(Mental state) Tired 1 2 3 4 5 Very good
Have you nodded off/slept a while?	No 1 2 3 4 5 Yes
Did you feel to control the feedback?	(Visual) No 1 2 3 4 5 Yes
	(Haptic) No 1 2 3 4 5 Yes
	(Multimodal) No 1 2 3 4 5 Yes
How easy was motor imagery?	Hard 1 2 3 4 5 Easy
After the motor imagery experiment	
Which type of feedback did you prefer?	0: v / 1: h / 2: v-h
How do you feel?	Anxious 1 2 3 4 5 Relaxed
	Bored 1 2 3 4 5 Excited
How was this experiment?	(Duration) Too long 1 2 3 4 5 Good
	(Timing) Too fast 1 2 3 4 5 Good
	(Enviroment) Poor 1 2 3 4 5 Good
	(System) Uncomfortable 1 2 3 4 5 Comfortable

detection of motor imagery. Hence, the classification accuracy associated with each experimental condition is calculated and compared to other conditions.

In data processing, baseline removal is firstly carried out. Specifically, a period of 100 ms before the cue is used. Then, the time-varying analysis is performed for all subjects, blocks and sessions by means of the cross-validation technique.

Subsequently, a permutation test is performed per each session, subject and block. The purpose is to validate the results obtained in the time-varying analysis evaluating how far these differed from the random classification. Hence, the labels associated with the left and right motor imagery tasks are randomly permuted and the time-varying analysis is repeated. The time-varying accuracy is calculated with the 2.00-s time window and 0.25-s shift by firstly exploiting the true labels. The mean of accuracies in cross-validation is taken into account for each window. The same time-varying accuracy is also calculated with the permuted labels. Finally, the comparison between the first results obtained and those from the permutation analysis is carried out using the non-parametric *Wilcoxon test* with the null hypothesis that the means of the two distributions are equal. Thus, rejecting the null hypothesis imply that the accuracy achieved with the true label is non-random. The significance level for the test is fixed at $\alpha = 5\,\%$.

Furthermore, based on the best 2 s time window in terms of classification accuracy, the *One-Way Analysis Of Variance* (*ANOVA*) is performed to compare the accuracy between blocks and sessions. The ANOVA is firstly conducted to analyze the difference between sessions per each subject and block and the difference between blocks per each subject and session. Then, by considering all subjects together, this analysis is done for each block to highlight differences between sessions and for each session to analyze differences between blocks. Before applying the ANOVA, inherent data assumptions are verified, i.e., data must be normally distributed and with equal variance (homoscedasticity). In case of normally distributed data samples with different variances, the Welch's correction is applied before using ANOVA. Instead, ANOVA could not be used when data were not normal, and a non-parametric test, namely the *Kruskal-Wallis test* is set in those cases.

11.2.2 Discussion

A representative example of the time-varying analysis plots generated from the original and randomly permuted data is reported in Fig. 11.13 for subject SB1. The mean classification accuracy and the associated standard deviation are calculated with a cross-validation across trials and by considering the shifting time window. The three different sessions are reported on rows, while the four different feedback conditions are reported on columns. Each plot has time in seconds on the x-axis and mean classification accuracy expressed in percentage on the y-axis. The black curves indicate the accuracy corresponding to the permuted labels, while the green line corresponds to the results reached using the true labels. As it would be expected, the two lines are indistinguishable up to the cue (at 2 s) in most cases. Then, it is noticeable that the two lines separate during the motor imagery.

At a first glance, it can be seen that the classification is more distant from random when neurofeedback is exploited. Moreover, better performance is also obtained in the last session, thus pointing out a training effect for the user. Similar results could be obtained with other subjects, though in some cases the classification is compatible with the chance level. The results of the *Wilcoxon test* associated with the permutation test demonstrated that, as a whole, four subjects in the first session and six in the second session achieved non-random accuracy when no feedback is provided. Then, five to seven subjects, out of eight, achieved non-random accuracy in the first session with feedback, with

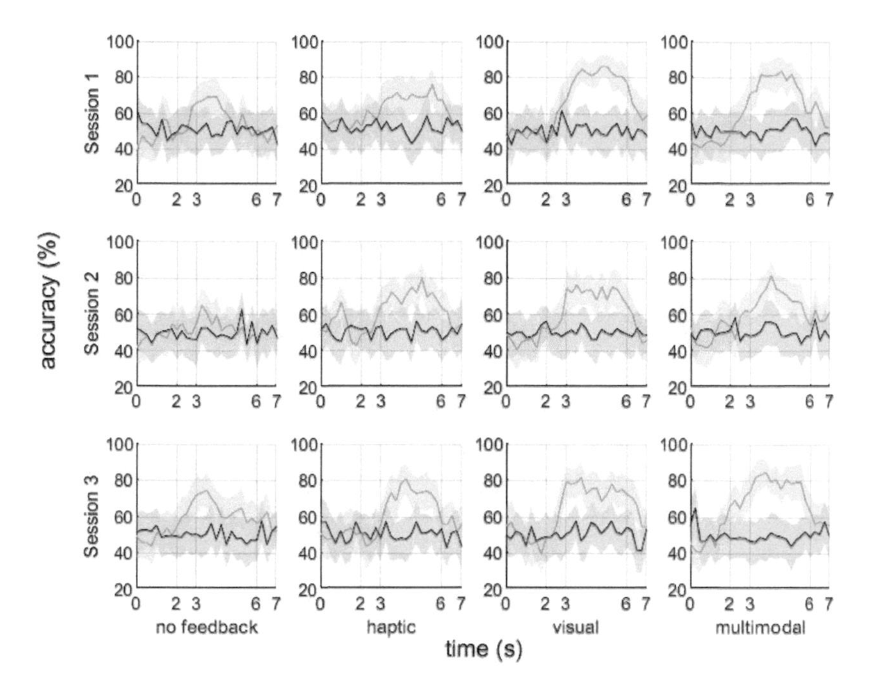

FIGURE 11.13 An example of the permutation test for the subject SB1 executed on the time course of classification accuracy in different experimental conditions. The green curves correspond to the accuracy calculated with true labels and the black curves correspond to the accuracy obtained with permuted labels.

no dominant feedback modality. In the second session, instead, at least six subjects achieved non-random accuracy.

The results are compatible with literature, since they highlight that a training effect subsists between sessions and that feedback is useful for increasing the performance. Notably, such results have been obtained with a wearable and portable system implemented with commercial hardware, which is a relatively novel trend for the BCI field.

Also the results from the third session (with only four subjects out of eight) point out an increase in overall performance. Nonetheless, mean classification accuracies appear compatible across the sessions and further experiments will be needed to properly reveal an improved detection across sessions. On the other hand, the improvement due to the neurofeedback is already evident. This is also shown in Fig. 11.14. In there, the time-varying classification accuracy associated with subject SB1 is compared to the one associated with the subject A03 from BCI competition IV. The Fig. 11.14(a) highlights that the trained subject A03 has substantially higher classification accuracy in the

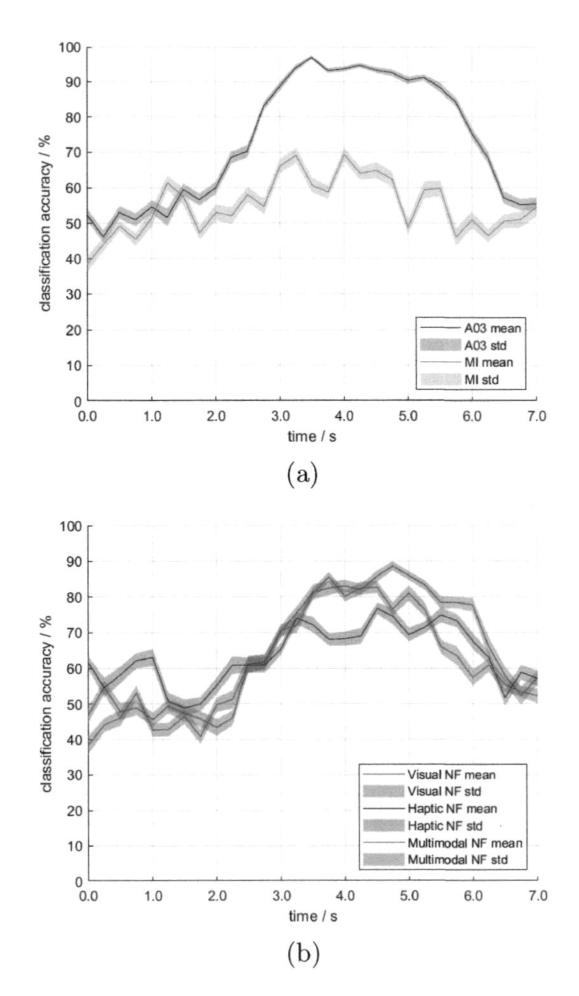

FIGURE 11.14 Time varying decoding accuracy associated with motor imagery. The classification accuracy of subject SB1 is compared to the one of a trained subject (A03 from BCI competition IV): no feedback (a) and with neurofeedback (b).

motor imagery period if compared to the performance of subject SB1. Then, Fig. 11.14(b) shows that the accuracies obtained with the neurofeedback become closer to the one of a trained subject.

Finally, the ANOVA executed as discussed above allowed to rigorously compare the classification accuracies in different conditions. As already mentioned, no statistically significant difference is revealed when considering the performance across sessions for all subjects. Moreover, the performance improvement with feedback is still not significant when all subjects are taken altogether. On the contrary, with the ANOVA executed subject-by-subject, the accuracy improvement is statistically significant for about half of the subjects between sessions, and the feedback results effective for six subjects out of eight.

In conclusion, neurofeedback resulted useful for most subjects in a daily-life BCI for improving motor imagery detection despite few channels being used, though future experiments are needed to further assess the system performance and better evaluate its limitations.

The wearability and portability of the discussed systems, as well as the usage of commercial hardware, can lead to easier implementation of motor rehabilitation. To improve the user-machine co-adaptation, future processing approaches should also involve Transfer Learning techniques.

Bibliography

[1] H. Berger, "Über das elektrenkephalogramm des menschen," *Archiv für psychiatrie und nervenkrankheiten*, vol. 87, no. 1, pp. 527–570, 1929.

[2] E. E. Green, A. M. Green, and E. D. Walters, "Voluntary control of internal states: Psychological and physiological," *The Journal of Transpersonal Psychology*, vol. 2, no. 1, p. 1, 1970.

[3] D. Rosenboom, "Method for producing sounds or light flashes with alpha brain waves for artistic purposes," *Leonardo*, pp. 141–145, 1972.

[4] J. J. Vidal, "Toward direct brain-computer communication," *Annual review of Biophysics and Bioengineering*, vol. 2, no. 1, pp. 157–180, 1973.

[5] J. J. Vidal, "Real-time detection of brain events in EEG," *Proceedings of the IEEE*, vol. 65, no. 5, pp. 633–641, 1977.

[6] J. R. Wolpaw, N. Birbaumer, W. J. Heetderks, D. J. McFarland, P. H. Peckham, G. Schalk, E. Donchin, L. A. Quatrano, C. J. Robinson, and T. M. Vaughan, "Brain-computer interface technology: a review of the first international meeting," *IEEE transactions on rehabilitation engineering*, vol. 8, no. 2, pp. 164–173, 2000.

[7] T. M. Vaughan, W. Heetderks, L. Trejo, W. Rymer, M. Weinrich, M. Moore, A. Kübler, B. Dobkin, N. Birbaumer, E. Donchin, *et al.*, "Brain-computer interface technology: a review of the second international meeting," 2003.

[8] F. Lotte, L. Bougrain, A. Cichocki, M. Clerc, M. Congedo, A. Rakotomamonjy, and F. Yger, "A review of classification algorithms for eeg-based brain–computer interfaces: a 10 year update," *Journal of neural engineering*, vol. 15, no. 3, p. 031005, 2018.

[9] Y.-H. Yu, S.-W. Lu, C.-H. Chuang, J.-T. King, C.-L. Chang, S.-A. Chen, S.-F. Chen, and C.-T. Lin, "An inflatable and wearable wireless system for making 32-channel electroencephalogram measurements," *IEEE Transactions on Neural Systems and Rehabilitation Engineering*, vol. 24, no. 7, pp. 806–813, 2016.

[10] R. Abiri, S. Borhani, E. W. Sellers, Y. Jiang, and X. Zhao, "A comprehensive review of eeg-based brain–computer interface paradigms," *Journal of neural engineering*, vol. 16, no. 1, p. 011001, 2019.

[11] P. Yuan, X. Gao, B. Allison, Y. Wang, G. Bin, and S. Gao, "A study of the existing problems of estimating the information transfer rate in online brain–computer interfaces," *Journal of neural engineering*, vol. 10, no. 2, p. 026014, 2013.

[12] M. Hamedi, S.-H. Salleh, and A. M. Noor, "Electroencephalographic motor imagery brain connectivity analysis for BCI: a review," *Neural computation*, vol. 28, no. 6, pp. 999–1041, 2016.

[13] M. Gomez-Rodriguez, J. Peters, J. Hill, B. Schölkopf, A. Gharabaghi, and M. Grosse-Wentrup, "Closing the sensorimotor loop: haptic feedback facilitates decoding of motor imagery," *Journal of neural engineering*, vol. 8, no. 3, p. 036005, 2011.

[14] A. R. Donati, S. Shokur, E. Morya, D. S. Campos, R. C. Moioli, C. M. Gitti, P. B. Augusto, S. Tripodi, C. G. Pires, G. A. Pereira, *et al.*, "Long-term training with a brain-machine interface-based gait protocol induces partial neurological recovery in paraplegic patients," *Scientific reports*, vol. 6, p. 30383, 2016.

[15] B. Kerous, F. Skola, and F. Liarokapis, "EEG-based bci and video games: a progress report," *Virtual Reality*, vol. 22, no. 2, pp. 119–135, 2018.

[16] Y.-T. Wang, Y. Wang, and T.-P. Jung, "A cell-phone-based brain–computer interface for communication in daily life," *Journal of neural engineering*, vol. 8, no. 2, p. 025018, 2011.

[17] "Bci competitions - berlin brain-computer interface." `http://www.bbci.de/competition/`. 2000 - 2008.

[18] "Cybathlon 2020 global edition." `https://cybathlon.ethz.ch/en`. November 13-14, 2020.

[19] Y. Li, J. Pan, J. Long, T. Yu, F. Wang, Z. Yu, and W. Wu, "Multimodal BCIs: target detection, multidimensional control, and awareness evaluation in patients with disorder of consciousness," *Proceedings of the IEEE*, vol. 104, no. 2, pp. 332–352, 2016.

[20] S. N. Abdulkader, A. Atia, and M.-S. M. Mostafa, "Brain computer interfacing: Applications and challenges," *Egyptian Informatics Journal*, vol. 16, no. 2, pp. 213–230, 2015.

[21] A. Waziri, J. Claassen, R. M. Stuart, H. Arif, J. M. Schmidt, S. A. Mayer, N. Badjatia, L. L. Kull, E. S. Connolly, R. G. Emerson, *et al.*, "Intracortical electroencephalography in acute brain injury," *Annals of Neurology: Official Journal of the American Neurological Association and the Child Neurology Society*, vol. 66, no. 3, pp. 366–377, 2009.

[22] O. David, J. M. Kilner, and K. J. Friston, "Mechanisms of evoked and induced responses in MEG/EEG," *Neuroimage*, vol. 31, no. 4, pp. 1580–1591, 2006.

[23] J. R. Wolpaw, N. Birbaumer, D. J. McFarland, G. Pfurtscheller, and T. M. Vaughan, "Brain–computer interfaces for communication and control," *Clinical neurophysiology*, vol. 113, no. 6, pp. 767–791, 2002.

[24] B. Z. Allison, D. J. McFarland, G. Schalk, S. D. Zheng, M. M. Jackson, and J. R. Wolpaw, "Towards an independent brain–computer interface using steady state visual evoked potentials," *Clinical neurophysiology*, vol. 119, no. 2, pp. 399–408, 2008.

[25] R. A. Ramadan and A. V. Vasilakos, "Brain computer interface: control signals review," *Neurocomputing*, vol. 223, pp. 26–44, 2017.

[26] T. O. Zander, C. Kothe, S. Jatzev, and M. Gaertner, "Enhancing human-computer interaction with input from active and passive brain-computer interfaces," in *Brain-computer interfaces*, pp. 181–199, Springer, 2010.

[27] D. Tan and A. Nijholt, "Brain-computer interfaces and human-computer interaction," in *Brain-Computer Interfaces*, pp. 3–19, Springer, 2010.

[28] M. Teplan *et al.*, "Fundamentals of EEG measurement," *Measurement science review*, vol. 2, no. 2, pp. 1–11, 2002.

[29] H. Hinrichs, M. Scholz, A. K. Baum, J. W. Kam, R. T. Knight, and H.-J. Heinze, "Comparison between a wireless dry electrode EEG system with a conventional wired wet electrode EEG system for clinical applications," *Scientific Reports*, vol. 10, no. 1, pp. 1–14, 2020.

[30] G. H. Klem, H. O. Lüders, H. H. Jasper, *et al.*, "The ten-twenty electrode system of the international federation," *Electroencephalogr Clin Neurophysiol*, vol. 52, no. 3, pp. 3–6, 1999.

[31] L. Angrisani, P. Arpaia, A. Esposito, and N. Moccaldi, "A wearable brain–computer interface instrument for augmented reality-based inspection in industry 4.0," *IEEE Transactions on Instrumentation and Measurement*, vol. 69, no. 4, pp. 1530–1539, 2019.

[32] M. Fatourechi, A. Bashashati, R. K. Ward, and G. E. Birch, "EMG and EOG artifacts in brain computer interface systems: A survey," *Clinical neurophysiology*, vol. 118, no. 3, pp. 480–494, 2007.

[33] G. James, D. Witten, T. Hastie, and R. Tibshirani, *An introduction to statistical learning*, vol. 112. Springer, 2013.

[34] K. Fukunaga, *Introduction to statistical pattern recognition.* Elsevier, 2013.

[35] P. Cunningham and S. J. Delany, "K-nearest neighbour classifiers-a tutorial," *ACM Computing Surveys (CSUR)*, vol. 54, no. 6, pp. 1–25, 2021.

[36] L. Rokach and O. Maimon, "Decision trees," in *Data mining and knowledge discovery handbook*, pp. 165–192, Springer, 2005.

[37] P. Probst, M. N. Wright, and A.-L. Boulesteix, "Hyperparameters and tuning strategies for random forest," *Wiley Interdisciplinary Reviews: Data Mining and Knowledge Discovery*, vol. 9, no. 3, p. e1301, 2019.

[38] P. Domingos and M. Pazzani, "On the optimality of the simple bayesian classifier under zero-one loss," *Machine learning*, vol. 29, no. 2-3, pp. 103–130, 1997.

[39] C. M. Bishop, *Pattern recognition and machine learning.* springer, 2006.

[40] W. S. Noble, "What is a support vector machine?," *Nature biotechnology*, vol. 24, no. 12, pp. 1565–1567, 2006.

[41] S. Arlot, A. Celisse, "A survey of cross-validation procedures for model selection," *Statistics surveys*, vol. 4, pp. 40–79, 2010.

[42] S. Federici and M. Scherer, *Assistive technology assessment handbook.* CRC Press, 2012.

[43] J. Malmivuo, R. Plonsey, *Bioelectromagnetism: principles and applications of bioelectric and biomagnetic fields.* Oxford University Press, USA, 1995.

[44] S. Musall, V. von Pföstl, A. Rauch, N. K. Logothetis, and K. Whittingstall, "Effects of neural synchrony on surface EEG," *Cerebral Cortex*, vol. 24, no. 4, pp. 1045–1053, 2014.

[45] S. Baillet, J. C. Mosher, and R. M. Leahy, "Electromagnetic brain mapping," *IEEE Signal processing magazine*, vol. 18, no. 6, pp. 14–30, 2001.

[46] D. I. Hoult and B. Bhakar, "NMR signal reception: Virtual photons and coherent spontaneous emission," *Concepts in Magnetic Resonance: An Educational Journal*, vol. 9, no. 5, pp. 277–297, 1997.

[47] A. Nijholt, D. Tan, G. Pfurtscheller, C. Brunner, J. d. R. Millán, B. Allison, B. Graimann, F. Popescu, B. Blankertz, and K.-R. Müller, "Brain-computer interfacing for intelligent systems," *IEEE intelligent systems*, vol. 23, no. 3, pp. 72–79, 2008.

[48] N. J. Hill, T. N. Lal, M. Schröder, T. Hinterberger, G. Widman, C. E. Elger, B. Schölkopf, and N. Birbaumer, "Classifying event-related desynchronization in EEG, ECOG and MEG signals," in *Joint Pattern Recognition Symposium*, pp. 404–413, Springer, 2006.

[49] L. F. Nicolas-Alonso and J. Gomez-Gil, "Brain computer interfaces, a review," *sensors*, vol. 12, no. 2, pp. 1211–1279, 2012.

[50] D. P. Subha, P. K. Joseph, R. Acharya, and C. M. Lim, "EEG signal analysis: a survey," *Journal of medical systems*, vol. 34, no. 2, pp. 195–212, 2010.

[51] G. Pfurtscheller, K. Pichler-Zalaudek, B. Ortmayr, J. Diez, *et al.*, "Post-movement beta synchronization in patients with parkinson's disease," *Journal of clinical neurophysiology*, vol. 15, no. 3, pp. 243–250, 1998.

[52] B. Blankertz, G. Curio, and K.-R. Müller, "Classifying single trial EEG: Towards brain computer interfacing," *Advances in neural information processing systems*, vol. 14, pp. 157–164, 2001.

[53] G. Pfurtscheller and F. L. Da Silva, "Event-related EEG/MEG synchronization and desynchronization: basic principles," *Clinical neurophysiology*, vol. 110, no. 11, pp. 1842–1857, 1999.

[54] S. S. Purkayastha, V. Jain, and H. Sardana, "Topical review: A review of various techniques used for measuring brain activity in brain computer interfaces," *Advance in Electronic and Electric Engineering*, vol. 4, pp. 513–522, 2014.

[55] Y. Wang, X. Gao, B. Hong, C. Jia, and S. Gao, "Brain-computer interfaces based on visual evoked potentials," *IEEE Engineering in medicine and biology magazine*, vol. 27, no. 5, pp. 64–71, 2008.

[56] G. Bin, X. Gao, Y. Wang, B. Hong, and S. Gao, "Vep-based brain-computer interfaces: time, frequency, and code modulations [research frontier]," *IEEE Computational Intelligence Magazine*, vol. 4, no. 4, pp. 22–26, 2009.

[57] D. J. Krusienski, E. W. Sellers, D. J. McFarland, T. M. Vaughan, and J. R. Wolpaw, "Toward enhanced p300 speller performance," *Journal of neuroscience methods*, vol. 167, no. 1, pp. 15–21, 2008.

[58] G. Pfurtscheller and C. Neuper, "Motor imagery and direct brain-computer communication," *Proceedings of the IEEE*, vol. 89, no. 7, pp. 1123–1134, 2001.

[59] C. Guger, H. Ramoser, and G. Pfurtscheller, "Real-time eeg analysis with subject-specific spatial patterns for a brain-computer interface (bci)," *IEEE transactions on rehabilitation engineering*, vol. 8, no. 4, pp. 447–456, 2000.

[60] O. Ltd, "Schematics of the EEG-SMT device for electroencephalography." `https://www.olimex.com/Products/EEG/OpenEEG/EEG-SMT/resources/EEG-SMT-SCHEMATIC-REV-B.pdf`.

[61] "Olimex EEG-SMT Website." https://www.olimex.com/Products/EEG/OpenEEG/EEG-SMT/open-source-hardware. Accessed: 2019-09-30.

[62] "ab medica s.p.a.." https://www.abmedica.it/.

[63] "Emotiv Epoc+ technical specifications." https://emotiv.gitbook.io/epoc-user-manual/introduction-1/technical_specifications.

[64] "Emotic Epoc+ Website." https://www.emotiv.com/epoc/. Accessed: 2019-09-30.

[65] "Neuroconcise ltd." https://www.neuroconcise.co.uk/.

[66] J. O. Bockris, B. E. Conway, E. Yeager, and R. E. White, *Comprehensive treatise of electrochemistry*, vol. 1. Springer, 1980.

[67] J. G. Webster, *Strumentazione biomedica: progetto ed applicazioni.* EdiSES, 2010.

[68] X. An and G. K. Stylios, "A hybrid textile electrode for electrocardiogram (ecg) measurement and motion tracking," *Materials*, vol. 11, no. 10, p. 1887, 2018.

[69] A. Cömert, M. Honkala, and J. Hyttinen, "Effect of pressure and padding on motion artifact of textile electrodes," *Biomedical engineering online*, vol. 12, no. 1, pp. 1–18, 2013.

[70] T. Alotaiby, F. E. A. El-Samie, S. A. Alshebeili, and I. Ahmad, "A review of channel selection algorithms for EEG signal processing," *EURASIP Journal on Advances in Signal Processing*, vol. 2015, no. 1, p. 66, 2015.

[71] H. Shan, H. Xu, S. Zhu, and B. He, "A novel channel selection method for optimal classification in different motor imagery BCI paradigms," *Biomedical engineering online*, vol. 14, no. 1, p. 93, 2015.

[72] C. Brunner, R. Leeb, G. Müller-Putz, A. Schlögl, and G. Pfurtscheller, "BCI competition 2008–graz data set a," *Institute for Knowledge Discovery (Laboratory of Brain-Computer Interfaces), Graz University of Technology*, vol. 16, pp. 1–6, 2008.

[73] K. K. Ang, Z. Y. Chin, C. Wang, C. Guan, and H. Zhang, "Filter bank common spatial pattern algorithm on BCI competition IV datasets 2a and 2b," *Frontiers in neuroscience*, vol. 6, p. 39, 2012.

[74] L. F. Nicolas-Alonso, R. Corralejo, J. Gomez-Pilar, D. Álvarez, and R. Hornero, "Adaptive stacked generalization for multiclass motor imagery-based brain computer interfaces," *IEEE Transactions on Neural Systems and Rehabilitation Engineering*, vol. 23, no. 4, pp. 702–712, 2015.

[75] P. Gaur, R. B. Pachori, H. Wang, and G. Prasad, "A multi-class EEG-based BCI classification using multivariate empirical mode decomposition based filtering and Riemannian geometry," *Expert Systems with Applications*, vol. 95, pp. 201–211, 2018.

[76] Y. Zhang, C. S. Nam, G. Zhou, J. Jin, X. Wang, and A. Cichocki, "Temporally constrained sparse group spatial patterns for motor imagery BCI," *IEEE transactions on cybernetics*, vol. 49, no. 9, pp. 3322–3332, 2018.

[77] A. Singh, S. Lal, and H. W. Guesgen, "Reduce calibration time in motor imagery using spatially regularized symmetric positives-definite matrices based classification," *Sensors*, vol. 19, no. 2, p. 379, 2019.

[78] B. E. Olivas-Padilla and M. I. Chacon-Murguia, "Classification of multiple motor imagery using deep convolutional neural networks and spatial filters," *Applied Soft Computing*, vol. 75, pp. 461–472, 2019.

[79] M. Arvaneh, C. Guan, K. K. Ang, and C. Quek, "Optimizing the channel selection and classification accuracy in EEG-based BCI," *IEEE Transactions on Biomedical Engineering*, vol. 58, no. 6, pp. 1865–1873, 2011.

[80] Y. Yang, S. Chevallier, J. Wiart, and I. Bloch, "Subject-specific time-frequency selection for multi-class motor imagery-based BCIs using few Laplacian EEG channels," *Biomedical Signal Processing and Control*, vol. 38, pp. 302–311, 2017.

[81] Y. Yang, I. Bloch, S. Chevallier, and J. Wiart, "Subject-specific channel selection using time information for motor imagery brain–computer interfaces," *Cognitive Computation*, vol. 8, no. 3, pp. 505–518, 2016.

[82] S. Ge, R. Wang, and D. Yu, "Classification of four-class motor imagery employing single-channel electroencephalography," *PloS one*, vol. 9, no. 6, p. e98019, 2014.

[83] L. Angrisani, P. Arpaia, F. Donnarumma, A. Esposito, N. Moccaldi, and M. Parvis, "Metrological performance of a single-channel Brain-Computer Interface based on Motor Imagery," in *2019 IEEE International Instrumentation and Measurement Technology Conference (I2MTC)*, pp. 1–5, IEEE, 2019.

[84] M. K. Ahirwal and M. R. Kose, "Audio-visual stimulation based emotion classification by correlated eeg channels," *Health and Technology*, vol. 10, no. 1, pp. 7–23, 2020.

[85] H. Xu, X. Wang, W. Li, H. Wang, and Q. Bi, "Research on EEG channel selection method for emotion recognition," in *2019 IEEE International Conference on Robotics and Biomimetics (ROBIO)*, pp. 2528–2535, IEEE, 2019.

[86] S. R. Sinha, L. R. Sullivan, D. Sabau, D. S. J. Orta, K. E. Dombrowski, J. J. Halford, A. J. Hani, F. W. Drislane, and M. M. Stecker, "American clinical neurophysiology society guideline 1: minimum technical requirements for performing clinical electroencephalography," *The Neurodiagnostic Journal*, vol. 56, no. 4, pp. 235–244, 2016.

[87] N. A. Badcock, K. A. Preece, B. de Wit, K. Glenn, N. Fieder, J. Thie, and G. McArthur, "Validation of the Emotiv EPOC EEG system for research quality auditory event-related potentials in children," *PeerJ*, vol. 3, p. e907, 2015.

[88] J. M. Rogers, S. J. Johnstone, A. Aminov, J. Donnelly, and P. H. Wilson, "Test-retest reliability of a single-channel, wireless EEG system," *International Journal of Psychophysiology*, vol. 106, pp. 87–96, 2016.

[89] M. Labecki, R. Kus, A. Brzozowska, T. Stacewicz, B. S. Bhattacharya, and P. Suffczynski, "Nonlinear origin of ssvep spectra—a combined experimental and modeling study," *Frontiers in computational neuroscience*, vol. 10, p. 129, 2016.

[90] IEEE Standards Association, "Neurotechnologies for Brain-Machine Interfacing." `https://standards.ieee.org/industry-connections/neurotechnologies-for-brain-machine-interfacing.html`.

[91] "Epson Moverio BT-200 Website." `https://www.olimex.com/Products/EEG/OpenEEG/EEG-SMT/resources/EEG-SMT.pdf`. Accessed: 2019-09-30.

[92] A. Cultrera, D. Corminboeuf, V. D'Elia, N. T. M. Tran, L. Callegaro, and M. Ortolano, "A new calibration setup for lock-in amplifiers in the low frequency range and its validation in a bilateral comparison," *Metrologia*, 2021.

[93] IEEE Standards Association, "IEEE 1057-2017 - IEEE Standard for Digitizing Waveform Recorders." `https://standards.ieee.org/standard/1057-2017.html`.

[94] Marko Neitola, "Four-Parameter Sinefit." `https://it.mathworks.com/matlabcentral/fileexchange/23214-four-parameter-sinefit`, 2021.

[95] P. Händel, "Amplitude estimation using IEEE-STD-1057 three-parameter sine wave fit: Statistical distribution, bias and variance," *Measurement*, vol. 43, no. 6, pp. 766–770, 2010.

[96] Joint Committee for Guides in Metrology (JCGM) - BIPM, "Evaluation of measurement data — Guide to the expression of uncertainty in measurement." `https://www.bipm.org/documents/20126/2071204/JCGM_100_2008_E.pdf/cb0ef43f-baa5-11cf-3f85-4dcd86f77bd6`, 2008.

[97] P. Arpaia, L. Callegaro, A. Cultrera, A. Esposito, and M. Ortolano, "Metrological characterization of consumer-grade equipment for wearable brain–computer interfaces and extended reality," *IEEE Transactions on Instrumentation and Measurement*, vol. 71, pp. 1–9, 2021.

[98] N. Bigdely-Shamlo, S. Makeig, and K. A. Robbins, "Preparing laboratory and real-world eeg data for large-scale analysis: a containerized approach," *Frontiers in neuroinformatics*, vol. 10, p. 7, 2016.

[99] V. Jayaram and A. Barachant, "Moabb: trustworthy algorithm benchmarking for bcis," *Journal of neural engineering*, vol. 15, no. 6, p. 066011, 2018.

[100] M. Ienca, P. Haselager, and E. J. Emanuel, "Brain leaks and consumer neurotechnology," *Nature biotechnology*, vol. 36, no. 9, pp. 805–810, 2018.

[101] M. Ienca and P. Haselager, "Hacking the brain: brain–computer interfacing technology and the ethics of neurosecurity," *Ethics and Information Technology*, vol. 18, no. 2, pp. 117–129, 2016.

[102] "Research Centre of European Private Law." `https://www.unisob.na.it/ateneo/c008.htm?vr=1`.

[103] M. Spüler, "A high-speed brain-computer interface (bci) using dry eeg electrodes," *PloS one*, vol. 12, no. 2, p. e0172400, 2017.

[104] A. Chabuda, P. Durka, and J. Żygierewicz, "High frequency ssvep-bci with hardware stimuli control and phase-synchronized comb filter," *IEEE Transactions on neural systems and rehabilitation engineering*, vol. 26, no. 2, pp. 344–352, 2017.

[105] N. Hill and B. Schölkopf, "An online brain–computer interface based on shifting attention to concurrent streams of auditory stimuli," *Journal of neural engineering*, vol. 9, no. 2, p. 026011, 2012.

[106] C. Breitwieser, V. Kaiser, C. Neuper, and G. R. Müller-Putz, "Stability and distribution of steady-state somatosensory evoked potentials elicited by vibro-tactile stimulation," *Medical & biological engineering & computing*, vol. 50, no. 4, pp. 347–357, 2012.

[107] S. Ajami, A. Mahnam, and V. Abootalebi, "Development of a practical high frequency brain–computer interface based on steady-state visual evoked potentials using a single channel of eeg," *Biocybernetics and Biomedical Engineering*, vol. 38, no. 1, pp. 106–114, 2018.

[108] H. Cecotti, "A self-paced and calibration-less ssvep-based brain–computer interface speller," *IEEE transactions on neural systems and rehabilitation engineering*, vol. 18, no. 2, pp. 127–133, 2010.

[109] M. Cheng, X. Gao, S. Gao, and D. Xu, "Design and implementation of a brain-computer interface with high transfer rates," *IEEE transactions on biomedical engineering*, vol. 49, no. 10, pp. 1181–1186, 2002.

[110] Y. Wang, R. Wang, X. Gao, B. Hong, and S. Gao, "A practical vep-based brain-computer interface," *IEEE Transactions on neural systems and rehabilitation engineering*, vol. 14, no. 2, pp. 234–240, 2006.

[111] I. Volosyak, F. Gembler, and P. Stawicki, "Age-related differences in SSVEP-based BCI performance," *Neurocomputing*, vol. 250, pp. 57–64, 2017.

[112] D. Lesenfants, D. Habbal, Z. Lugo, M. Lebeau, P. Horki, E. Amico, C. Pokorny, F. Gomez, A. Soddu, G. Müller-Putz, *et al.*, "An independent ssvep-based brain–computer interface in locked-in syndrome," *Journal of neural engineering*, vol. 11, no. 3, p. 035002, 2014.

[113] H. Si-Mohammed, F. A. Sanz, G. Casiez, N. Roussel, and A. Lécuyer, "Brain-computer interfaces and augmented reality: A state of the art," in *Graz Brain-Computer Interface Conference*, 2017.

[114] V. Mihajlović, B. Grundlehner, R. Vullers, and J. Penders, "Wearable, wireless eeg solutions in daily life applications: what are we missing?," *IEEE journal of biomedical and health informatics*, vol. 19, no. 1, pp. 6–21, 2014.

[115] J. Minguillon, M. A. Lopez-Gordo, and F. Pelayo, "Trends in EEG-BCI for daily-life: Requirements for artifact removal," *Biomedical Signal Processing and Control*, vol. 31, pp. 407–418, 2017.

[116] M. Wang, R. Li, R. Zhang, G. Li, and D. Zhang, "A wearable SSVEP-based bci system for quadcopter control using head-mounted device," *IEEE Access*, vol. 6, pp. 26789–26798, 2018.

[117] H. Si-Mohammed, J. Petit, C. Jeunet, F. Argelaguet, F. Spindler, A. Évain, N. Roussel, G. Casiez, and A. Lécuyer, "Towards BCI-based interfaces for augmented reality: Feasibility, design and evaluation," *IEEE transactions on visualization and computer graphics*, 2018.

[118] T.-H. Nguyen and W.-Y. Chung, "A Single-Channel SSVEP-Based BCI Speller Using Deep Learning," *IEEE Access*, vol. 7, pp. 1752–1763, 2019.

[119] R. Leeb, F. Lee, C. Keinrath, R. Scherer, H. Bischof, and G. Pfurtscheller, "Brain–computer communication: motivation, aim, and impact of exploring a virtual apartment," *IEEE Transactions on Neural Systems and Rehabilitation Engineering*, vol. 15, no. 4, pp. 473–482, 2007.

[120] "GTec Website." `https://www.gtec.at/it/`. Accessed: 2019-09-30.

[121] "NeurpSkt Website." `http://neurosky.com/`. Accessed: 2019-09-30.

[122] "bitbrain Website." `https://www.bitbrain.com/`. Accessed: 2019-09-30.

[123] "Open Source Brain-Computer Interfaces." `https://openbci.com/`.

[124] T. Radüntz, "Signal quality evaluation of emerging EEG devices," *Frontiers in physiology*, vol. 9, p. 98, 2018.

[125] W. Klonowski, "Everything you wanted to ask about eeg but were afraid to get the right answer," *Nonlinear biomedical physics*, vol. 3, no. 1, pp. 1–5, 2009.

[126] G. Li, S. Wang, and Y. Y. Duan, "Towards gel-free electrodes: A systematic study of electrode-skin impedance," *Sensors and Actuators B: Chemical*, vol. 241, pp. 1244–1255, 2017.

[127] L. Casal and G. La Mura, "Skin-electrode impedance measurement during ECG acquisition: method's validation," in *Journal of Physics: Conference Series*, vol. 705, p. 012006, IOP Publishing, 2016.

[128] Z. Zhao, K. Ivanov, L. Lubich, O. M. Omisore, Z. Mei, N. Fu, J. Chen, and L. Wang, "Signal Quality and Electrode-Skin Impedance Evaluation in the Context of Wearable Electroencephalographic Systems," in *2018 40th Annual International Conference of the IEEE Engineering in Medicine and Biology Society (EMBC)*, pp. 4965–4968, IEEE, 2018.

[129] F. Grosselin, X. Navarro-Sune, A. Vozzi, K. Pandremmenou, F. De Vico Fallani, Y. Attal, and M. Chavez, "Quality assessment of single-channel EEG for wearable devices," *Sensors*, vol. 19, no. 3, p. 601, 2019.

[130] B. Hu, H. Peng, Q. Zhao, B. Hu, D. Majoe, F. Zheng, and P. Moore, "Signal quality assessment model for wearable EEG sensor on prediction of mental stress," *IEEE transactions on nanobioscience*, vol. 14, no. 5, pp. 553–561, 2015.

[131] M. Sahu, S. Mohdiwale, N. Khoriya, Y. Upadhyay, A. Verma, and S. Singh, "Eeg artifact removal techniques: A comparative study," in *International Conference on Innovative Computing and Communications*, pp. 395–403, Springer, 2020.

[132] S. Blum, N. S. Jacobsen, M. G. Bleichner, and S. Debener, "A Riemannian modification of artifact subspace reconstruction for EEG artifact handling," *Frontiers in human neuroscience*, vol. 13, p. 141, 2019.

[133] D. Chatzopoulos, C. Bermejo, Z. Huang, and P. Hui, "Mobile augmented reality survey: From where we are to where we go," *Ieee Access*, vol. 5, pp. 6917–6950, 2017.

[134] J. Grubert, Y. Itoh, K. Moser, and J. E. Swan, "A survey of calibration methods for optical see-through head-mounted displays," *IEEE transactions on visualization and computer graphics*, vol. 24, no. 9, pp. 2649–2662, 2017.

[135] M. Klemm, F. Seebacher, and H. Hoppe, "High accuracy pixel-wise spatial calibration of optical see-through glasses," *Computers & Graphics*, vol. 64, pp. 51–61, 2017.

[136] X. Zhao, C. Liu, Z. Xu, L. Zhang, and R. Zhang, "SSVEP stimulus layout effect on accuracy of brain-computer interfaces in augmented reality glasses," *IEEE Access*, vol. 8, pp. 5990–5998, 2020.

[137] "hololens Website." `https://www.microsoft.com/it-it/hololens`. Accessed: 2019-09-30.

[138] "Epson Moverio BT-200 Website." `https://www.epson.it/products/see-through-mobile-viewer/moverio-bt-200`. Accessed: 2019-09-30.

[139] A. Duszyk, M. Bierzyńska, Z. Radzikowska, P. Milanowski, R. Kuś, P. Suffczyński, M. Michalska, M. Łabęcki, P. Zwoliński, and P. Durka, "Towards an optimization of stimulus parameters for brain-computer interfaces based on steady state visual evoked potentials," *Plos one*, vol. 9, no. 11, p. e112099, 2014.

[140] P. Arpaia, L. Duraccio, N. Moccaldi, and S. Rossi, "Wearable brain–computer interface instrumentation for robot-based rehabilitation by augmented reality," *IEEE Transactions on Instrumentation and Measurement*, vol. 69, no. 9, pp. 6362–6371, 2020.

[141] S. Kanoga, M. Nakanishi, and Y. Mitsukura, "Assessing the effects of voluntary and involuntary eyeblinks in independent components of electroencephalogram," *Neurocomputing*, vol. 193, pp. 20–32, 2016.

[142] L. Angrisani, P. Arpaia, E. De Benedetto, A. Esposito, N. Moccaldi, and M. Parvis, "Brain-computer interfaces for daily-life applications: a five-year experience report," in *2021 IEEE International Instrumentation and Measurement Technology Conference (I2MTC)*, pp. 1–6, IEEE, 2021.

[143] P. Refaeilzadeh, L. Tang, and H. Liu, "Cross-validation," *Encyclopedia of database systems*, vol. 5, pp. 532–538, 2009.

[144] H. Hsu and P. A. Lachenbruch, "Paired t test," *Wiley StatsRef: statistics reference online*, 2014.

[145] P. Stawicki, F. Gembler, A. Rezeika, and I. Volosyak, "A novel hybrid mental spelling application based on eye tracking and SSVEP-based BCI," *Brain sciences*, vol. 7, no. 4, p. 35, 2017.

[146] F. P. Borges, D. M. Garcia, *et al.*, "Distribution of spontaneous inter-blink interval in repeated measurements with and without topical ocular anesthesia," *Arquivos brasileiros de oftalmologia*, vol. 73, no. 4, pp. 329–332, 2010.

[147] J. R. Wolpaw, H. Ramoser, D. J. McFarland, and G. Pfurtscheller, "Eeg-based communication: improved accuracy by response verification," *IEEE transactions on Rehabilitation Engineering*, vol. 6, no. 3, pp. 326–333, 1998.

[148] X. Xing, Y. Wang, W. Pei, X. Guo, Z. Liu, F. Wang, G. Ming, H. Zhao, Q. Gui, and H. Chen, "A high-speed ssvep-based bci using dry EEG electrodes," *Scientific reports*, vol. 8, no. 1, pp. 1–10, 2018.

[149] A. Luo and T. J. Sullivan, "A user-friendly ssvep-based brain–computer interface using a time-domain classifier," *Journal of neural engineering*, vol. 7, no. 2, p. 026010, 2010.

[150] C.-C. Lo, T.-Y. Chien, J.-S. Pan, and B.-S. Lin, "Novel non-contact control system for medical healthcare of disabled patients," *IEEE Access*, vol. 4, pp. 5687–5694, 2016.

[151] M. Rüßmann, M. Lorenz, P. Gerbert, M. Waldner, J. Justus, P. Engel, and M. Harnisch, "Industry 4.0: The future of productivity and growth in manufacturing industries," *Boston Consulting Group*, vol. 9, 2015.

[152] P. Arpaia, A. Covino, L. Cristaldi, M. Frosolone, L. Gargiulo, F. Mancino, F. Mantile, and N. Moccaldi, "A systematic review on feature extraction in electroencephalography-based diagnostics and therapy in attention deficit hyperactivity disorder," *Sensors*, vol. 22, no. 13, p. 4934, 2022.

[153] National Collaborating Centre for Mental Health, "Attention deficit hyperactivity disorder: diagnosis and management of adhd in children, young people and adults," British Psychological Society, 2018.

[154] A. C. Pierno, M. Mari, D. Lusher, and U. Castiello, "Robotic movement elicits visuomotor priming in children with autism," *Neuropsychologia*, vol. 46, no. 2, pp. 448–454, 2008.

[155] B. H. Cho, J.-M. Lee, J. Ku, D. P. Jang, J. Kim, I.-Y. Kim, J.-H. Lee, and S. I. Kim, "Attention enhancement system using virtual reality and eeg biofeedback," in *Proceedings IEEE Virtual Reality 2002*, pp. 156–163, IEEE, 2002.

[156] C. G. Lim, T. S. Lee, C. Guan, D. S. S. Fung, Y. Zhao, S. S. W. Teng, H. Zhang, and K. R. R. Krishnan, "A brain-computer interface based attention training program for treating attention deficit hyperactivity disorder," *PloS one*, vol. 7, no. 10, p. e46692, 2012.

[157] G. Pires, M. Torres, N. Casaleiro, U. Nunes, and M. Castelo-Branco, "Playing tetris with non-invasive bci," in *2011 IEEE 1st International Conference on Serious Games and Applications for Health (SeGAH)*, pp. 1–6, IEEE, 2011.

[158] D. Z. Blandón, J. E. Munoz, D. S. Lopez, and O. H. Gallo, "Influence of a BCI neurofeedback videogame in children with ADHD. quantifying the brain activity through an EEG signal processing dedicated toolbox," in *2016 IEEE 11th Colombian Computing Conference (CCC)*, pp. 1–8, IEEE, 2016.

[159] D. A. Rohani, H. B. Sorensen, and S. Puthusserypady, "Brain-computer interface using p300 and virtual reality: A gaming approach for treating adhd," in *2014 36th Annual International Conference of the IEEE Engineering in Medicine and Biology Society*, pp. 3606–3609, IEEE, 2014.

[160] L. Jiang, C. Guan, H. Zhang, C. Wang, and B. Jiang, "Brain computer interface based 3D game for attention training and rehabilitation," in *2011 6th IEEE Conference on Industrial Electronics and Applications*, pp. 124–127, IEEE, 2011.

[161] L. J. Buxbaum, M. Ferraro, T. Veramonti, A. Farne, J. Whyte, E. Ladavas, F. Frassinetti, and H. Coslett, "Hemispatial neglect: Subtypes, neuroanatomy, and disability," *Neurology*, vol. 62, no. 5, pp. 749–756, 2004.

[162] M. Sohlberg, "Theory and remediation of attention disorders," *Introduction to Cognitive Rehabilitation Theory & Practice*, pp. 110–135, 1989.

[163] M. Noam, N. Mor, S. Arjen, R. T. Knight, and A. Perry, "Behavioral and eeg measures show no amplifying effects of shared attention on attention or memory," *Scientific Reports (Nature Publisher Group)*, vol. 10, no. 1, 2020.

[164] B. L. Trommer, J.-A. B. Hoeppner, R. Lorber, and K. J. Armstrong, "The go—no-go paradigm in attention deficit disorder," *Annals of neurology*, vol. 24, no. 5, pp. 610–614, 1988.

[165] T. J. McDermott, A. I. Wiesman, A. L. Proskovec, E. Heinrichs-Graham, and T. W. Wilson, "Spatiotemporal oscillatory dynamics of visual selective attention during a flanker task," *Neuroimage*, vol. 156, pp. 277–285, 2017.

[166] C. A. Riccio, C. R. Reynolds, and P. A. Lowe, *Clinical applications of continuous performance tests: Measuring attention and impulsive responding in children and adults.* John Wiley & Sons Inc, 2001.

[167] C. Bench, C. Frith, P. Grasby, K. Friston, E. Paulesu, R. Frackowiak, and R. J. Dolan, "Investigations of the functional anatomy of attention

using the stroop test," *Neuropsychologia*, vol. 31, no. 9, pp. 907–922, 1993.

[168] M. Leclercq and P. Zimmermann, *Applied neuropsychology of attention: theory, diagnosis and rehabilitation.* Psychology Press, 2004.

[169] J. H. Kim, C. M. Kim, E.-S. Jung, and M.-S. Yim, "Biosignal based attention monitoring to support nuclear operator safety-relevant tasks," *Frontiers in computational neuroscience*, vol. 14, p. 111, 2020.

[170] K. P. Tee, C. Guan, K. K. Ang, K. S. Phua, C. Wang, and H. Zhang, "Augmenting cognitive processes in robot-assisted motor rehabilitation," in *2008 2nd IEEE RAS & EMBS International Conference on Biomedical Robotics and Biomechatronics*, pp. 698–703, IEEE, 2008.

[171] K. K. Ang and C. Guan, "Brain-computer interface in stroke rehabilitation," *Journal of Computing Science and Engineering*, vol. 7, no. 2, pp. 139–146, 2013.

[172] S. C. Cramer, M. Sur, B. H. Dobkin, C. O'brien, T. D. Sanger, J. Q. Trojanowski, J. M. Rumsey, R. Hicks, J. Cameron, D. Chen, *et al.*, "Harnessing neuroplasticity for clinical applications," *Brain*, vol. 134, no. 6, pp. 1591–1609, 2011.

[173] G. Loheswaran, M. S. Barr, R. Zomorrodi, T. K. Rajji, D. M. Blumberger, B. Le Foll, and Z. J. Daskalakis, "Impairment of neuroplasticity in the dorsolateral prefrontal cortex by alcohol," *Scientific reports*, vol. 7, no. 1, pp. 1–8, 2017.

[174] P. R. Kleinginna and A. M. Kleinginna, "A categorized list of emotion definitions, with suggestions for a consensual definition," *Motivation and emotion*, vol. 5, no. 4, pp. 345–379, 1981.

[175] P. Ekman, "Emotions revealed," *Bmj*, vol. 328, no. Suppl S5, 2004.

[176] A. Mehrabian and J. A. Russell, "The basic emotional impact of environments," *Perceptual and motor skills*, vol. 38, no. 1, pp. 283–301, 1974.

[177] J. C. Borod, B. A. Cicero, L. K. Obler, J. Welkowitz, H. M. Erhan, C. Santschi, I. S. Grunwald, R. M. Agosti, and J. R. Whalen, "Right hemisphere emotional perception: evidence across multiple channels.," *Neuropsychology*, vol. 12, no. 3, p. 446, 1998.

[178] G. L. Ahern and G. E. Schwartz, "Differential lateralization for positive versus negative emotion," *Neuropsychologia*, vol. 17, no. 6, pp. 693–698, 1979.

[179] R. J. Davidson, P. Ekman, C. D. Saron, J. A. Senulis, and W. V. Friesen, "Approach-withdrawal and cerebral asymmetry: emotional expression and brain physiology: I.," *Journal of personality and social psychology*, vol. 58, no. 2, p. 330, 1990.

[180] S. K. Sutton and R. J. Davidson, "Prefrontal brain asymmetry: A biological substrate of the behavioral approach and inhibition systems," *Psychological science*, vol. 8, no. 3, pp. 204–210, 1997.

[181] M. Golan, Y. Cohen, and G. Singer, "A framework for operator–workstation interaction in industry 4.0," *International Journal of Production Research*, vol. 58, no. 8, pp. 2421–2432, 2020.

[182] Y. Liu, O. Sourina, and M. K. Nguyen, "Real-time EEG-based emotion recognition and its applications," in *Transactions on computational science XII*, pp. 256–277, Springer, 2011.

[183] C. A. Pop, R. Simut, S. Pintea, J. Saldien, A. Rusu, D. David, J. Vanderfaeillie, D. Lefeber, and B. Vanderborght, "Can the social robot probo help children with autism to identify situation-based emotions? a series of single case experiments," *International Journal of Humanoid Robotics*, vol. 10, no. 03, p. 1350025, 2013.

[184] C. Jones and J. Sutherland, "Acoustic emotion recognition for affective computer gaming," in *Affect and emotion in human-computer interaction*, pp. 209–219, Springer, 2008.

[185] M. Poole and M. Warner, *The IEBM handbook of human resource management.* International Thomson Business, 1998.

[186] G. Matthews, L. Joyner, K. Gilliland, S. Campbell, S. Falconer, and J. Huggins, "Validation of a comprehensive stress state questionnaire: Towards a state big three," *Personality psychology in Europe*, vol. 7, pp. 335–350, 1999.

[187] N. Sharma and T. Gedeon, "Objective measures, sensors and computational techniques for stress recognition and classification: A survey," *Computer methods and programs in biomedicine*, vol. 108, no. 3, pp. 1287–1301, 2012.

[188] J. Fischer, A. Calame, A. Dettling, H. Zeier, and S. Fanconi, "Objectifying psychomental stress in the workplace–an example," *International archives of occupational and environmental health*, vol. 73, no. 1, pp. S46–S52, 2000.

[189] D. S. Goldstein, "Stress-induced activation of the sympathetic nervous system," *Bailliere's clinical endocrinology and metabolism*, vol. 1, no. 2, pp. 253–278, 1987.

[190] J. Minguillon, E. Perez, M. Lopez-Gordo, F. Pelayo, and M. Sanchez-Carrion, "Portable System for Real-Time Detection of Stress Level," *Sensors*, vol. 18, no. 8, p. 2504, 2018.

[191] A. Mohammed and L. Wang, "Brainwaves driven human-robot collaborative assembly," *CIRP annals*, vol. 67, no. 1, pp. 13–16, 2018.

[192] M. Choi, G. Koo, M. Seo, and S. W. Kim, "Wearable device-based system to monitor a driver's stress, fatigue, and drowsiness," *IEEE Transactions on Instrumentation and Measurement*, vol. 67, no. 3, pp. 634–645, 2017.

[193] A. Lay-Ekuakille, P. Vergallo, G. Griffo, F. Conversano, S. Casciaro, S. Urooj, V. Bhateja, and A. Trabacca, "Entropy index in quantitative EEG measurement for diagnosis accuracy," *IEEE Transactions on Instrumentation and Measurement*, vol. 63, no. 6, pp. 1440–1450, 2013.

[194] B. Wallace, F. Knoefel, R. Goubran, R. A. L. Zunini, Z. Ren, and A. Maccosham, "Eeg/erp: within episodic assessment framework for cognition," *IEEE Transactions on Instrumentation and Measurement*, vol. 66, no. 10, pp. 2525–2534, 2017.

[195] H. Jebelli, S. Hwang, and S. Lee, "EEG-based workers' stress recognition at construction sites," *Automation in Construction*, vol. 93, pp. 315–324, 2018.

[196] L.-D. Liao, S.-L. Wu, C.-H. Liou, S.-W. Lu, S.-A. Chen, S.-F. Chen, L.-W. Ko, and C.-T. Lin, "A novel 16-channel wireless system for electroencephalography measurements with dry spring-loaded sensors," *IEEE Transactions on Instrumentation and Measurement*, vol. 63, no. 6, pp. 1545–1555, 2014.

[197] Y.-C. Chen, B.-S. Lin, and J.-S. Pan, "Novel noncontact dry electrode with adaptive mechanical design for measuring EEG in a hairy site," *IEEE Transactions on Instrumentation and Measurement*, vol. 64, no. 12, pp. 3361–3368, 2015.

[198] C. Setz, B. Arnrich, J. Schumm, R. La Marca, G. Tröster, and U. Ehlert, "Discriminating stress from cognitive load using a wearable EDA device," *IEEE Transactions on information technology in biomedicine*, vol. 14, no. 2, pp. 410–417, 2009.

[199] W. A. Kahn, "Psychological conditions of personal engagement and disengagement at work," *Academy of management journal*, vol. 33, no. 4, pp. 692–724, 1990.

[200] S. Barello, S. Triberti, G. Graffigna, C. Libreri, S. Serino, J. Hibbard, and G. Riva, "ehealth for patient engagement: a systematic review," *Frontiers in psychology*, vol. 6, p. 2013, 2016.

[201] A. H. Lequerica and K. Kortte, "Therapeutic engagement: a proposed model of engagement in medical rehabilitation," *American journal of physical medicine & rehabilitation*, vol. 89, no. 5, pp. 415–422, 2010.

[202] E. L. Deci and R. M. Ryan, "Conceptualizations of intrinsic motivation and self-determination," in *Intrinsic motivation and self-determination in human behavior*, pp. 11–40, Springer, 1985.

[203] P. Gibbons, *Scaffolding language, scaffolding learning*. Portsmouth, NH: Heinemann, 2002.

[204] G. King, L. A. Chiarello, R. Ideishi, R. D'Arrigo, E. Smart, J. Ziviani, and M. Pinto, "The nature, value, and experience of engagement in pediatric rehabilitation: perspectives of youth, caregivers, and service providers," *Developmental neurorehabilitation*, vol. 23, no. 1, pp. 18–30, 2020.

[205] A. Lutz, H. A. Slagter, J. D. Dunne, and R. J. Davidson, "Attention regulation and monitoring in meditation," *Trends in cognitive sciences*, vol. 12, no. 4, pp. 163–169, 2008.

[206] J. P. Connell and J. G. Wellborn, "Competence, autonomy, and relatedness: A motivational analysis of self-system processes.," in *Cultural processes in child development: The Minnesota symposia on child psychology*, vol. 23, pp. 43–78, Psychology Press, 1991.

[207] L. P. Anjarichert, K. Gross, K. Schuster, and S. Jeschke, "Learning 4.0: Virtual immersive engineering education," *Digit. Univ*, vol. 2, p. 51, 2016.

[208] D. Janssen, C. Tummel, A. Richert, and I. Isenhardt, "Virtual environments in higher education-immersion as a key construct for learning 4.0.," *iJAC*, vol. 9, no. 2, pp. 20–26, 2016.

[209] L. M. Andreessen, P. Gerjets, D. Meurers, and T. O. Zander, "Toward neuroadaptive support technologies for improving digital reading: A passive bci-based assessment of mental workload imposed by text difficulty and presentation speed during reading," *User Modeling and User-Adapted Interaction*, vol. 31, no. 1, pp. 75–104, 2021.

[210] F. Paas, J. E. Tuovinen, H. Tabbers, and P. W. Van Gerven, "Cognitive load measurement as a means to advance cognitive load theory," *Educational psychologist*, vol. 38, no. 1, pp. 63–71, 2003.

[211] S. McLeod, "Jean piaget's theory of cognitive development," *Simply Psychology*, pp. 1–9, 2018.

[212] J. A. Kleim and T. A. Jones, "Principles of experience-dependent neural plasticity: implications for rehabilitation after brain damage," 2008.

[213] R. VanDeWeghe, *Engaged learning.* Corwin Press, 2009.

[214] H. Jang, J. Reeve, and E. L. Deci, "Engaging students in learning activities: It is not autonomy support or structure but autonomy support and structure.," *Journal of educational psychology*, vol. 102, no. 3, p. 588, 2010.

[215] K. Schweizer and H. Moosbrugger, "Attention and working memory as predictors of intelligence," *Intelligence*, vol. 32, no. 4, pp. 329–347, 2004.

[216] N. Mrachacz-Kersting, A. J. T. Stevenson, S. Aliakbaryhosseinabadi, A. C. Lundgaard, H. R. Jørgensen, K. Severinsen, and D. Farina, "An associative brain-computer-interface for acute stroke patients," in *Converging Clinical and Engineering Research on Neurorehabilitation II*, pp. 841–845, Springer, 2017.

[217] J. Yang, W. Li, S. Wang, J. Lu, and L. Zou, "Classification of children with attention deficit hyperactivity disorder using PCA and k-nearest neighbors during interference control task," in *Advances in Cognitive Neurodynamics (V)*, pp. 447–453, Springer, 2016.

[218] Y. Akimoto, T. Nozawa, A. Kanno, M. Ihara, T. Goto, T. Ogawa, T. Kambara, M. Sugiura, E. Okumura, and R. Kawashima, "High-gamma activity in an attention network predicts individual differences in elderly adults' behavioral performance," *Neuroimage*, vol. 100, pp. 290–300, 2014.

[219] L. da Silva-Sauer, L. Valero-Aguayo, A. de la Torre-Luque, R. Ron-Angevin, and S. Varona-Moya, "Concentration on performance with p300-based bci systems: A matter of interface features," *Applied ergonomics*, vol. 52, pp. 325–332, 2016.

[220] S. Aliakbaryhosseinabadi, E. N. Kamavuako, N. Jiang, D. Farina, and N. Mrachacz-Kersting, "Classification of eeg signals to identify variations in attention during motor task execution," *Journal of neuroscience methods*, vol. 284, pp. 27–34, 2017.

[221] B. Hamadicharef, H. Zhang, C. Guan, C. Wang, K. S. Phua, K. P. Tee, and K. K. Ang, "Learning eeg-based spectral-spatial patterns for attention level measurement," in *2009 IEEE International Symposium on Circuits and Systems*, pp. 1465–1468, IEEE, 2009.

[222] J. M. Antelis, L. Montesano, X. Giralt, A. Casals, and J. Minguez, "Detection of movements with attention or distraction to the motor task during robot-assisted passive movements of the upper limb," in *2012 Annual International Conference of the IEEE Engineering in Medicine and Biology Society*, pp. 6410–6413, IEEE, 2012.

[223] S. Paradiso, D. L. Johnson, N. C. Andreasen, D. S. O'Leary, G. L. Watkins, L. L. Boles Ponto, and R. D. Hichwa, "Cerebral blood flow changes associated with attribution of emotional valence to pleasant, unpleasant, and neutral visual stimuli in a pet study of normal subjects," *American Journal of Psychiatry*, vol. 156, no. 10, pp. 1618–1629, 1999.

[224] J. Perdiz, G. Pires, and U. J. Nunes, "Emotional state detection based on emg and eog biosignals: A short survey," in *2017 IEEE 5th Portuguese Meeting on Bioengineering (ENBENG)*, pp. 1–4, IEEE, 2017.

[225] M. Benovoy, J. R. Cooperstock, and J. Deitcher, "Biosignals analysis and its application in a performance setting," in *Proceedings of the International Conference on Bio-Inspired Systems and Signal Processing*, pp. 253–258, 2008.

[226] S. S. Stevens, "The direct estimation of sensory magnitudes: Loudness," *The American journal of psychology*, vol. 69, no. 1, pp. 1–25, 1956.

[227] P. De Bièvre, "The 2012 international vocabulary of metrology:"vim"," *Accreditation and Quality Assurance*, vol. 17, no. 2, pp. 231–232, 2012.

[228] P. E. Ekman and R. J. Davidson, *The nature of emotion: Fundamental questions*. Oxford University Press, 1994.

[229] P. J. Lang, "International affective picture system (iaps): Affective ratings of pictures and instruction manual," *Technical report*, 2005.

[230] R. Jenke, A. Peer, and M. Buss, "Feature extraction and selection for emotion recognition from EEG," *IEEE Transactions on Affective computing*, vol. 5, no. 3, pp. 327–339, 2014.

[231] H. A. Demaree, D. E. Everhart, E. A. Youngstrom, and D. W. Harrison, "Brain lateralization of emotional processing: historical roots and a future incorporating "dominance"," *Behavioral and cognitive neuroscience reviews*, vol. 4, no. 1, pp. 3–20, 2005.

[232] J. A. Coan and J. J. Allen, "The state and trait nature of frontal eeg asymmetry in emotion.," 2003.

[233] R. J. Davidson, "Hemispheric asymmetry and emotion," *Approaches to emotion*, vol. 2, pp. 39–57, 1984.

[234] A. Bechara, A. R. Damasio, H. Damasio, S. W. Anderson, *et al.*, "Insensitivity to future consequences following damage to human prefrontal cortex," *Cognition*, vol. 50, pp. 1–3, 1994.

[235] D. Hagemann, E. Naumann, G. Becker, S. Maier, and D. Bartussek, "Frontal brain asymmetry and affective style: A conceptual replication," *Psychophysiology*, vol. 35, no. 4, pp. 372–388, 1998.

[236] J. A. Coan and J. J. Allen, "Frontal EEG asymmetry as a moderator and mediator of emotion," *Biological psychology*, vol. 67, no. 1-2, pp. 7–50, 2004.

[237] J. Wolpaw and E. W. Wolpaw, *Brain-computer interfaces: principles and practice.* OUP USA, 2012.

[238] H. Zeng, Z. Wu, J. Zhang, C. Yang, H. Zhang, G. Dai, and W. Kong, "EEG emotion classification using an improved sincnet-based deep learning model," *Brain sciences*, vol. 9, no. 11, p. 326, 2019.

[239] Y. Luo, G. Wu, S. Qiu, S. Yang, W. Li, and Y. Bi, "EEG-based emotion classification using deep neural network and sparse autoencoder," *Frontiers in Systems Neuroscience*, vol. 14, p. 43, 2020.

[240] Y. Luo, Q. Fu, J. Xie, Y. Qin, G. Wu, J. Liu, F. Jiang, Y. Cao, and X. Ding, "EEG-based emotion classification using spiking neural networks," *IEEE Access*, vol. 8, pp. 46007–46016, 2020.

[241] F. Wang, S. Wu, W. Zhang, Z. Xu, Y. Zhang, C. Wu, and S. Coleman, "Emotion recognition with convolutional neural network and EEG-based efdms," *Neuropsychologia*, p. 107506, 2020.

[242] Y. Cimtay and E. Ekmekcioglu, "Investigating the use of pretrained convolutional neural network on cross-subject and cross-dataset EEG emotion recognition," *Sensors*, vol. 20, no. 7, p. 2034, 2020.

[243] T. Song, W. Zheng, P. Song, and Z. Cui, "EEG emotion recognition using dynamical graph convolutional neural networks," *IEEE Transactions on Affective Computing*, 2018.

[244] J. Chen, D. Jiang, and Y. Zhang, "A hierarchical bidirectional gru model with attention for EEG-based emotion classification," *IEEE Access*, vol. 7, pp. 118530–118540, 2019.

[245] M. Z. Soroush, K. Maghooli, S. K. Setarehdan, and A. M. Nasrabadi, "Emotion classification through nonlinear EEG analysis using machine learning methods," *International Clinical Neuroscience Journal*, vol. 5, no. 4, p. 135, 2018.

[246] H. Ullah, M. Uzair, A. Mahmood, M. Ullah, S. D. Khan, and F. A. Cheikh, "Internal emotion classification using EEG signal with sparse discriminative ensemble," *IEEE Access*, vol. 7, pp. 40144–40153, 2019.

[247] D. D. Chakladar and S. Chakraborty, "EEG based emotion classification using "correlation based subset selection"," *Biologically inspired cognitive architectures*, vol. 24, pp. 98–106, 2018.

[248] H. A. Gonzalez, S. Muzaffar, J. Yoo, and I. M. Elfadel, "Biocnn: A hardware inference engine for EEG-based emotion detection," *IEEE Access*, vol. 8, pp. 140896–140914, 2020.

[249] X. Xing, Z. Li, T. Xu, L. Shu, B. Hu, and X. Xu, "Sae+ lstm: A new framework for emotion recognition from multi-channel EEG," *Frontiers in Neurorobotics*, vol. 13, p. 37, 2019.

[250] F. Yang, X. Zhao, W. Jiang, P. Gao, and G. Liu, "Cross-subject emotion recognition using multi-method fusion from high-dimensional features," *Frontiers in Computational Neuroscience*, vol. 13, p. 53, 2019.

[251] Y. Liu, Y. Ding, C. Li, J. Cheng, R. Song, F. Wan, and X. Chen, "Multi-channel EEG-based emotion recognition via a multi-level features guided capsule network," *Computers in Biology and Medicine*, vol. 123, p. 103927, 2020.

[252] H. Cui, A. Liu, X. Zhang, X. Chen, K. Wang, and X. Chen, "EEG-based emotion recognition using an end-to-end regional-asymmetric convolutional neural network," *Knowledge-Based Systems*, vol. 205, p. 106243, 2020.

[253] D. J. Hemanth, "EEG signal based modified kohonen neural networks for classification of human mental emotions," *Journal of Artificial Intelligence and Systems*, vol. 2, pp. 1–13, 2020.

[254] K. Guo, R. Chai, H. Candra, Y. Guo, R. Song, H. Nguyen, and S. Su, "A hybrid fuzzy cognitive map/support vector machine approach for EEG-based emotion classification using compressed sensing," *International Journal of Fuzzy Systems*, vol. 21, no. 1, pp. 263–273, 2019.

[255] W.-L. Zheng and B.-L. Lu, "Investigating critical frequency bands and channels for EEG-based emotion recognition with deep neural networks," *IEEE Transactions on Autonomous Mental Development*, vol. 7, no. 3, pp. 162–175, 2015.

[256] "Seed-dataset." `https://bcmi.sjtu.edu.cn/home/seed/seed.html`, 2021.

[257] S. Koelstra, C. Muhl, M. Soleymani, J.-S. Lee, A. Yazdani, T. Ebrahimi, T. Pun, A. Nijholt, and I. Patras, "Deap: A database for emotion analysis; using physiological signals," *IEEE transactions on affective computing*, vol. 3, no. 1, pp. 18–31, 2011.

[258] "Deap-dataset." `http://www.eecs.qmul.ac.uk/mmv/datasets/deap/readme.html`, 2021.

[259] "BioSemi Website." `https://www.biosemi.com/`. Accessed: 2019-09-30.

[260] S. Katsigiannis and N. Ramzan, "Dreamer: A database for emotion recognition through EEG and ecg signals from wireless low-cost off-the-shelf devices," *IEEE journal of biomedical and health informatics*, vol. 22, no. 1, pp. 98–107, 2017.

[261] "Dreamer-dataset." `https://zenodo.org/record/546113#.YMNN6NUzbIU`, 2021.

[262] R.-N. Duan, J.-Y. Zhu, and B.-L. Lu, "Differential entropy feature for EEG-based emotion classification," in *6th International IEEE/EMBS Conference on Neural Engineering (NER)*, pp. 81–84, IEEE, 2013.

[263] Q. Gao, C.-h. Wang, Z. Wang, X.-l. Song, E.-z. Dong, and Y. Song, "EEG based emotion recognition using fusion feature extraction method," *Multimedia Tools and Applications*, vol. 79, no. 37, pp. 27057–27074, 2020.

[264] R. M. Mehmood and H. J. Lee, "EEG based emotion recognition from human brain using hjorth parameters and svm," *International Journal of Bio-Science and Bio-Technology*, vol. 7, no. 3, pp. 23–32, 2015.

[265] E. S. Dan-Glauser and K. R. Scherer, "The geneva affective picture database (gaped): a new 730-picture database focusing on valence and normative significance," *Behavior research methods*, vol. 43, no. 2, p. 468, 2011.

[266] S. Taran and V. Bajaj, "Emotion recognition from single-channel EEG signals using a two-stage correlation and instantaneous frequency-based filtering method," *Computer Methods and Programs in Biomedicine*, vol. 173, pp. 157–165, 2019.

[267] M. Ogino and Y. Mitsukura, "A mobile application for estimating emotional valence using a single-channel EEG device," in *2018 57th Annual Conference of the Society of Instrument and Control Engineers of Japan (SICE)*, pp. 1043–1048, IEEE, 2018.

[268] N. Jatupaiboon, S. Pan-ngum, and P. Israsena, "Emotion classification using minimal EEG channels and frequency bands," in *The 2013 10th International Joint Conference on Computer Science and Software Engineering (JCSSE)*, pp. 21–24, IEEE, 2013.

[269] A. Jalilifard, E. B. Pizzolato, and M. K. Islam, "Emotion classification using single-channel scalp-EEG recording," in *2016 38th Annual International Conference of the IEEE Engineering in Medicine and Biology Society (EMBC)*, pp. 845–849, IEEE, 2016.

[270] P. C. Petrantonakis and L. J. Hadjileontiadis, "Emotion recognition from EEG using higher order crossings," *IEEE Transactions on information Technology in Biomedicine*, vol. 14, no. 2, pp. 186–197, 2009.

[271] P. Pandey and K. Seeja, "Emotional state recognition with EEG signals using subject independent approach," in *Data Science and Big Data Analytics*, pp. 117–124, Springer, 2019.

[272] A. Q.-X. Ang, Y. Q. Yeong, and W. Wee, "Emotion classification from EEG signals using time-frequency-DWT features and ANN," *Journal of Computer and Communications*, vol. 5, no. 3, pp. 75–79, 2017.

[273] C.-J. Yang, N. Fahier, C.-Y. He, W.-C. Li, and W.-C. Fang, "An ai-edge platform with multimodal wearable physiological signals monitoring sensors for affective computing applications," in *2020 IEEE International Symposium on Circuits and Systems (ISCAS)*, pp. 1–5, IEEE, 2020.

[274] J. Marín-Morales, J. L. Higuera-Trujillo, A. Greco, J. Guixeres, C. Llinares, E. P. Scilingo, M. Alcañiz, and G. Valenza, "Affective computing in virtual reality: emotion recognition from brain and heartbeat dynamics using wearable sensors," *Scientific reports*, vol. 8, no. 1, pp. 1–15, 2018.

[275] Y. Wei, Y. Wu, and J. Tudor, "A real-time wearable emotion detection headband based on EEG measurement," *Sensors and Actuators A: Physical*, vol. 263, pp. 614–621, 2017.

[276] A. Apicella, P. Arpaia, G. Mastrati, and N. Moccaldi, "Eeg-based detection of emotional valence towards a reproducible measurement of emotions," *Scientific Reports*, vol. 11, no. 1, pp. 1–16, 2021.

[277] A.-M. Brouwer, M. A. Neerincx, V. Kallen, L. van der Leer, and M. ten Brinke, "EEG alpha asymmetry, heart rate variability and cortisol in response to virtual reality induced stress," *Journal of Cybertherapy & Rehabilitation*, vol. 4, no. 1, pp. 21–34, 2011.

[278] GR. Suresh, "Development of four stress levels in group stroop colour word test using HRV analysis.," *Biomedical Research (0970-938X)*, vol. 28, no. 1, 2017.

[279] I. Papousek, E. M. Weiss, G. Schulter, A. Fink, E. M. Reiser, and H. K. Lackner, "Prefrontal EEG alpha asymmetry changes while observing disaster happening to other people: cardiac correlates and prediction of emotional impact," *Biological psychology*, vol. 103, pp. 184–194, 2014.

[280] R. N. Goodman, J. C. Rietschel, L.-C. Lo, M. E. Costanzo, and B. D. Hatfield, "Stress, emotion regulation and cognitive performance: The predictive contributions of trait and state relative frontal EEG alpha asymmetry," *International Journal of Psychophysiology*, vol. 87, no. 2, pp. 115–123, 2013.

[281] J. Mladenović, J. Mattout, and F. Lotte, "A generic framework for adaptive EEG-based BCI training and operation," *arXiv preprint arXiv:1707.07935*, 2017.

[282] F. Lotte and C. Jeunet, "Online classification accuracy is a poor metric to study mental imagery-based bci user learning: an experimental demonstration and new metrics," 2017.

[283] D. J. Krusienski, M. Grosse-Wentrup, F. Galán, D. Coyle, K. J. Miller, E. Forney, and C. W. Anderson, "Critical issues in state-of-the-art brain–computer interface signal processing," *Journal of neural engineering*, vol. 8, no. 2, p. 025002, 2011.

[284] F. Lotte, "Signal processing approaches to minimize or suppress calibration time in oscillatory activity-based brain–computer interfaces," *Proceedings of the IEEE*, vol. 103, no. 6, pp. 871–890, 2015.

[285] M. Krauledat, M. Schröder, B. Blankertz, and K.-R. Müller, "Reducing calibration time for brain-computer interfaces: A clustering approach," in *Advances in Neural Information Processing Systems*, pp. 753–760, 2007.

[286] H. Woehrle, M. M. Krell, S. Straube, S. K. Kim, E. A. Kirchner, and F. Kirchner, "An adaptive spatial filter for user-independent single trial detection of event-related potentials," *IEEE Transactions on Biomedical Engineering*, vol. 62, no. 7, pp. 1696–1705, 2015.

[287] X. Chen, H. Peng, F. Yu, and K. Wang, "Independent vector analysis applied to remove muscle artifacts in EEG data," *IEEE Transactions on Instrumentation and Measurement*, vol. 66, no. 7, pp. 1770–1779, 2017.

[288] K. Takahashi, "Remarks on emotion recognition from multi-modal bio potential signals," *The Japanese Journal of Ergonomics*, vol. 41, no. 4, pp. 248–253, 2005.

[289] M. Zanetti, T. Mizumoto, L. Faes, A. Fornaser, M. De Cecco, L. Maule, M. Valente, and G. Nollo, "Multilevel assessment of mental stress via network physiology paradigm using consumer wearable devices," *Journal of Ambient Intelligence and Humanized Computing*, pp. 1–10, 2019.

[290] S. A. Hosseini and M. A. Khalilzadeh, "Emotional stress recognition system using EEG and psychophysiological signals: Using new labelling process of EEG signals in emotional stress state," in *2010 international conference on biomedical engineering and computer science*, pp. 1–6, IEEE, 2010.

[291] P. Zhang, X. Wang, X. Li, and P. Dai, "EEG feature selection based on weighted-normalized mutual information for mental fatigue classification," in *2016 IEEE International Instrumentation and Measurement Technology Conference Proceedings*, pp. 1–6, IEEE, 2016.

[292] F. Lotte, M. Congedo, A. Lécuyer, F. Lamarche, and B. Arnaldi, "A review of classification algorithms for EEG-based brain–computer interfaces," *Journal of neural engineering*, vol. 4, no. 2, p. R1, 2007.

[293] A. Schlögl, C. Vidaurre, and K.-R. Müller, "Adaptive methods in BCI research-an introductory tutorial," in *Brain-Computer Interfaces*, pp. 331–355, Springer, 2009.

[294] G. Liu, D. Zhang, J. Meng, G. Huang, and X. Zhu, "Unsupervised adaptation of electroencephalogram signal processing based on fuzzy C-means algorithm," *International Journal of Adaptive Control and Signal Processing*, vol. 26, no. 6, pp. 482–495, 2012.

[295] X. Hou, Y. Liu, O. Sourina, Y. R. E. Tan, L. Wang, and W. Mueller-Wittig, "EEG based stress monitoring," in *2015 IEEE International Conference on Systems, Man, and Cybernetics*, pp. 3110–3115, IEEE, 2015.

[296] S. Mühlbacher-Karrer, A. H. Mosa, L.-M. Faller, M. Ali, R. Hamid, H. Zangl, and K. Kyamakya, "A driver state detection system—combining a capacitive hand detection sensor with physiological sensors," *IEEE Transactions on Instrumentation and Measurement*, vol. 66, no. 4, pp. 624–636, 2017.

[297] Z. Gao, S. Li, Q. Cai, W. Dang, Y. Yang, C. Mu, and P. Hui, "Relative wavelet entropy complex network for improving EEG-based fatigue driving classification," *IEEE Transactions on Instrumentation and Measurement*, no. 99, pp. 1–7, 2018.

[298] G. Jun and K. G. Smitha, "EEG based stress level identification," in *2016 IEEE International Conference on Systems, Man, and Cybernetics (SMC)*, pp. 003270–003274, IEEE, 2016.

[299] A. Secerbegovic, S. Ibric, J. Nisic, N. Suljanovic, and A. Mujcic, "Mental workload vs. stress differentiation using single-channel EEG," in *CMBE-BIH 2017*, pp. 511–515, Springer, 2017.

[300] D. Bian, J. W. Wade, A. Swanson, Z. Warren, and N. Sarkar, "Physiology-based affect recognition during driving in virtual environment for autism intervention.," in *PhyCS*, pp. 137–145, 2015.

[301] J. W. Ahn, Y. Ku, and H. C. Kim, "A Novel Wearable EEG and ECG Recording System for Stress Assessment," *Sensors*, vol. 19, no. 9, p. 1991, 2019.

[302] A. Wigfield, J. T. Guthrie, K. C. Perencevich, A. Taboada, S. L. Klauda, A. McRae, and P. Barbosa, "Role of reading engagement in mediating effects of reading comprehension instruction on reading outcomes," *Psychology in the Schools*, vol. 45, no. 5, pp. 432–445, 2008.

[303] S. Helme and D. Clarke, "Identifying cognitive engagement in the mathematics classroom," *Mathematics Education Research Journal*, vol. 13, no. 2, pp. 133–153, 2001.

[304] P.-S. D. Chen, A. D. Lambert, and K. R. Guidry, "Engaging online learners: The impact of web-based learning technology on college student engagement," *Computers & Education*, vol. 54, no. 4, pp. 1222–1232, 2010.

[305] S. Jaafar, N. S. Awaludin, and N. S. Bakar in *E-proceeding of the Conference on Management and Muamalah*, pp. 128–135, 2014.

[306] L. Angrisani, P. Arpaia, F. Donnarumma, A. Esposito, M. Frosolone, G. Improta, N. Moccaldi, A. Natalizio, and M. Parvis, "Instrumentation for motor imagery-based brain computer interfaces relying on dry electrodes: a functional analysis," in *2020 IEEE International Instrumentation and Measurement Technology Conference (I2MTC)*, pp. 1–6, IEEE, 2020.

[307] A. Apicella, P. Arpaia, M. Frosolone, and N. Moccaldi, "High-wearable EEG-based distraction detection in motor rehabilitation," *Scientific Reports*, vol. 11, no. 1, pp. 1–9, 2021.

[308] M. S. Benlamine, A. Dufresne, M. H. Beauchamp, and C. Frasson, "Bargain: behavioral affective rule-based games adaptation interface–towards emotionally intelligent games: application on a virtual reality environment for socio-moral development," *User Modeling and User-Adapted Interaction*, pp. 1–35, 2021.

[309] X.-W. Wang, D. Nie, and B.-L. Lu, "Emotional state classification from EEG data using machine learning approach," *Neurocomputing*, vol. 129, pp. 94–106, 2014.

[310] M. Soleymani, S. Asghari-Esfeden, Y. Fu, and M. Pantic, "Analysis of eeg signals and facial expressions for continuous emotion detection," *IEEE Transactions on Affective Computing*, vol. 7, no. 1, pp. 17–28, 2015.

[311] N. Zhuang, Y. Zeng, L. Tong, C. Zhang, H. Zhang, and B. Yan, "Emotion recognition from EEG signals using multidimensional information in emd domain," *BioMed research international*, vol. 2017, 2017.

[312] I. Jraidi, M. Chaouachi, and C. Frasson, "A hierarchical probabilistic framework for recognizing learners' interaction experience trends and emotions," *Advances in Human-Computer Interaction*, vol. 2014, 2014.

[313] P. Aricò, G. Borghini, G. Di Flumeri, A. Colosimo, S. Pozzi, and F. Babiloni, "A passive brain–computer interface application for the mental workload assessment on professional air traffic controllers during realistic air traffic control tasks," *Progress in brain research*, vol. 228, pp. 295–328, 2016.

[314] S. Wang, J. Gwizdka, and W. A. Chaovalitwongse, "Using wireless eeg signals to assess memory workload in the n-back task," *IEEE Transactions on Human-Machine Systems*, vol. 46, no. 3, pp. 424–435, 2015.

[315] J. H. Hibbard, E. R. Mahoney, J. Stockard, and M. Tusler, "Development and testing of a short form of the patient activation measure," *Health services research*, vol. 40, no. 6p1, pp. 1918–1930, 2005.

[316] G. Graffigna, S. Barello, A. Bonanomi, and E. Lozza, "Measuring patient engagement: development and psychometric properties of the patient health engagement (phe) scale," *Frontiers in psychology*, vol. 6, p. 274, 2015.

[317] G. King, L. A. Chiarello, L. Thompson, M. J. McLarnon, E. Smart, J. Ziviani, and M. Pinto, "Development of an observational measure of therapy engagement for pediatric rehabilitation," *Disability and Rehabilitation*, vol. 41, no. 1, pp. 86–97, 2019.

[318] E. Dell'Aquila, G. Maggi, D. Conti, and S. Rossi, "A preparatory study for measuring engagement in pediatric virtual and robotics rehabilitation settings," in *Companion of the 2020 ACM/IEEE International Conference on Human-Robot Interaction*, pp. 183–185, 2020.

[319] C. Ranti, W. Jones, A. Klin, and S. Shultz, "Blink rate patterns provide a reliable measure of individual engagement with scene content," *Scientific Reports*, vol. 10, no. 1, pp. 1–10, 2020.

[320] G. H. Gendolla, "Self-relevance of performance, task difficulty, and task engagement assessed as cardiovascular response," *Motivation and Emotion*, vol. 23, no. 1, pp. 45–66, 1999.

[321] A. R. Harrivel, D. H. Weissman, D. C. Noll, and S. J. Peltier, "Monitoring attentional state with fnirs," *Frontiers in human neuroscience*, vol. 7, p. 861, 2013.

[322] E. T. Esfahani, S. Pareek, P. Chembrammel, M. Ghobadi, and T. Kesavadas, "Adaptation of rehabilitation system based on user's mental engagement," in *ASME 2015 International Design Engineering Technical Conferences and Computers and Information in Engineering Conference*, American Society of Mechanical Engineers Digital Collection, 2015.

[323] N. Kumar and K. P. Michmizos, "Machine learning for motor learning: Eeg-based continuous assessment of cognitive engagement for adaptive rehabilitation robots," in *2020 8th IEEE RAS/EMBS International Conference for Biomedical Robotics and Biomechatronics (BioRob)*, pp. 521–526, 2020.

[324] W. Park, G. H. Kwon, D.-H. Kim, Y.-H. Kim, S.-P. Kim, and L. Kim, "Assessment of cognitive engagement in stroke patients from single-trial eeg during motor rehabilitation," *IEEE Transactions on Neural Systems and Rehabilitation Engineering*, vol. 23, no. 3, pp. 351–362, 2014.

[325] X. Dang, R. Wei, and G. Li, "An efficient movement and mental classification for children with autism based on motion and EEG features," *Journal of Ambient Intelligence and Humanized Computing*, vol. 8, no. 6, pp. 907–912, 2017.

[326] L. Angrisani, P. Arpaia, D. Casinelli, and N. Moccaldi, "A single-channel ssvep-based instrument with off-the-shelf components for trainingless brain-computer interfaces," *IEEE Transactions on Instrumentation and Measurement*, 2018.

[327] A. Apicella, P. Arpaia, M. Frosolone, G. Improta, N. Moccaldi, and A. Pollastro, "EEG-based measurement system for monitoring student engagement in learning 4.0," *Scientific Reports*, vol. 12, no. 1, pp. 1–13, 2022.

[328] A. Delorme, T. Sejnowski, and S. Makeig, "Enhanced detection of artifacts in eeg data using higher-order statistics and independent component analysis," *Neuroimage*, vol. 34, no. 4, pp. 1443–1449, 2007.

[329] H. Abdi and L. J. Williams, "Principal component analysis," *Wiley interdisciplinary reviews: computational statistics*, vol. 2, no. 4, pp. 433–459, 2010.

[330] I. Jolliffe, *Principal component analysis.* Springer, 2011.

[331] J. Asensio-Cubero, J. Q. Gan, and R. Palaniappan, "Multiresolution analysis over graphs for a motor imagery based online BCI game," *Computers in biology and medicine*, vol. 68, pp. 21–26, 2016.

[332] W.-L. Zheng, Y.-Q. Zhang, J.-Y. Zhu, and B.-L. Lu, "Transfer components between subjects for EEG-based emotion recognition," in *2015 international conference on affective computing and intelligent interaction (ACII)*, pp. 917–922, IEEE, 2015.

[333] S. J. Pan and Q. Yang, "A survey on transfer learning," *IEEE Transactions on knowledge and data engineering*, vol. 22, no. 10, pp. 1345–1359, 2009.

[334] S. J. Pan, I. W. Tsang, J. T. Kwok, and Q. Yang, "Domain adaptation via transfer component analysis," *IEEE Transactions on Neural Networks*, vol. 22, no. 2, pp. 199–210, 2010.

[335] N. V. Chawla, K. W. Bowyer, L. O. Hall, and W. P. Kegelmeyer, "Smote: synthetic minority over-sampling technique," *Journal of artificial intelligence research*, vol. 16, pp. 321–357, 2002.

[336] H. Han, W.-Y. Wang, and B.-H. Mao, "Borderline-smote: a new over-sampling method in imbalanced data sets learning," in *International conference on intelligent computing*, pp. 878–887, Springer, 2005.

[337] Haibo He, Yang Bai, E. A. Garcia, and Shutao Li, "Adasyn: Adaptive synthetic sampling approach for imbalanced learning," in *2008 IEEE International Joint Conference on Neural Networks (IEEE World Congress on Computational Intelligence)*, pp. 1322–1328, 2008.

[338] H. M. Nguyen, E. W. Cooper, and K. Kamei, "Borderline over-sampling for imbalanced data classification," *Int. J. Knowl. Eng. Soft Data Paradigm.*, vol. 3, p. 4–21, Apr. 2011.

[339] G. Douzas, F. Bacao, and F. Last, "Improving imbalanced learning through a heuristic oversampling method based on k-means and smote," *Information Sciences*, vol. 465, p. 1–20, Oct 2018.

[340] J. B. MacQueen, "Some methods for classification and analysis of multivariate observations," in *Proc. of the fifth Berkeley Symposium on Mathematical Statistics and Probability* (L. M. L. Cam and J. Neyman, eds.), vol. 1, pp. 281–297, University of California Press, 1967.

[341] J. Pereira, A. I. Sburlea, and G. R. Müller-Putz, "Eeg patterns of self-paced movement imaginations towards externally-cued and internally-selected targets," *Scientific reports*, vol. 8, no. 1, pp. 1–15, 2018.

[342] P. Ye, X. Wu, D. Gao, H. Liang, J. Wang, S. Deng, N. Xu, J. She, and J. Chen, "Comparison of dp3 signals evoked by comfortable 3d images and 2d images—an event-related potential study using an oddball task," *Scientific reports*, vol. 7, p. 43110, 2017.

[343] J. Polich and C. Margala, "P300 and probability: comparison of oddball and single-stimulus paradigms," *International Journal of Psychophysiology*, vol. 25, no. 2, pp. 169–176, 1997.

[344] S. A. Huettel and G. McCarthy, "What is odd in the oddball task?: Prefrontal cortex is activated by dynamic changes in response strategy," *Neuropsychologia*, vol. 42, no. 3, pp. 379–386, 2004.

[345] "Texasinstrument-ads1298." `https://www.ti.com/lit/ds/symlink/ads1296r.pdf`, 2020-02-28.

[346] S. Varma and R. Simon, "Bias in error estimation when using cross-validation for model selection," *BMC bioinformatics*, vol. 7, no. 1, p. 91, 2006.

[347] H. Cases, "Neurology 2021 market analysis," *Int J Collab Res Intern Med Public Health*, vol. 13, no. 1, p. 1043, 2021.

[348] K. Kroenke and R. L. Spitzer, "The phq-9: a new depression diagnostic and severity measure," *Psychiatric annals*, vol. 32, no. 9, pp. 509–515, 2002.

[349] L. M. Braunstein, J. J. Gross, and K. N. Ochsner, "Explicit and implicit emotion regulation: a multi-level framework," *Social cognitive and affective neuroscience*, vol. 12, no. 10, pp. 1545–1557, 2017.

[350] B. Kurdi, S. Lozano, and M. R. Banaji, "Introducing the open affective standardized image set (oasis)," *Behavior research methods*, vol. 49, no. 2, pp. 457–470, 2017.

[351] T. Radüntz, J. Scouten, O. Hochmuth, and B. Meffert, "Eeg artifact elimination by extraction of ica-component features using image processing algorithms," *Journal of neuroscience methods*, vol. 243, pp. 84–93, 2015.

[352] T. Hastie, R. Tibshirani, and J. Friedman, *The elements of statistical learning: data mining, inference, and prediction.* Springer Science & Business Media, 2009.

[353] C. Cortes and V. Vapnik, "Support-vector networks," *Machine learning*, vol. 20, no. 3, pp. 273–297, 1995.

[354] M. Bentlemsan, E.-T. Zemouri, D. Bouchaffra, B. Yahya-Zoubir, and K. Ferroudji, "Random forest and filter bank common spatial patterns for eeg-based motor imagery classification," in *2014 5th International conference on intelligent systems, modelling and simulation*, pp. 235–238, IEEE, 2014.

[355] B. Siciliano and O. Khatib, *Springer handbook of robotics.* Springer, 2016.

[356] P. Renaud and J.-P. Blondin, "The stress of stroop performance: Physiological and emotional responses to color–word interference, task pacing, and pacing speed," *International Journal of Psychophysiology*, vol. 27, no. 2, pp. 87–97, 1997.

[357] C. D. Spielberger, "Manual for the State-Trait Anxiety Inventory STAI (form Y)(" self-evaluation questionnaire")," 1983.

[358] J. Blascovich and J. Tomaka, "Measures of self-esteem," *Measures of personality and social psychological attitudes*, vol. 1, pp. 115–160, 1991.

[359] A. Celisse and S. Robin, "Nonparametric density estimation by exact leave-p-out cross-validation," *Computational Statistics & Data Analysis*, vol. 52, no. 5, pp. 2350–2368, 2008.

[360] A. Gaume, G. Dreyfus, and F.-B. Vialatte, "A cognitive brain–computer interface monitoring sustained attentional variations during a continuous task," *Cognitive neurodynamics*, vol. 13, no. 3, pp. 257–269, 2019.

[361] R. Panda, R. Malheiro, and R. P. Paiva, "Novel audio features for music emotion recognition," *IEEE Transactions on Affective Computing*, vol. 11, no. 4, pp. 614–626, 2018.

[362] J. A. Russell, "A circumplex model of affect.," *Journal of personality and social psychology*, vol. 39, no. 6, p. 1161, 1980.

[363] A. Delorme and S. Makeig, "EEGLAB: an open source toolbox for analysis of single-trial eeg dynamics including independent component analysis," *Journal of neuroscience methods*, vol. 134, no. 1, pp. 9–21, 2004.

[364] A. T. Pope, E. H. Bogart, and D. S. Bartolome, "Biocybernetic system evaluates indices of operator engagement in automated task," *Biological psychology*, vol. 40, no. 1-2, pp. 187–195, 1995.

[365] A. Apicella, P. Arpaia, S. Giugliano, G. Mastrati, and N. Moccaldi, "High-wearable eeg-based transducer for engagement detection in pediatric rehabilitation," *Brain-Computer Interfaces*, pp. 1–11, 2021.

[366] R. Chanubol, P. Wongphaet, N. Chavanich, C. Werner, S. Hesse, A. Bardeleben, and J. Merholz, "A randomized controlled trial of cognitive sensory motor training therapy on the recovery of arm function in acute stroke patients," *Clinical Rehabilitation*, vol. 26, no. 12, pp. 1096–1104, 2012. PMID: 22649162.

[367] S. W. Kozlowski and K. Hattrup, "A disagreement about within-group agreement: Disentangling issues of consistency versus consensus.," *Journal of applied psychology*, vol. 77, no. 2, p. 161, 1992.

[368] M. S. Myslobodsky and J. Bar-Ziv, "Locations of occipital eeg electrodes verified by computed tomography," *Electroencephalography and clinical neurophysiology*, vol. 72, no. 4, pp. 362–366, 1989.

[369] A. Mandel and S. Keller, "Stress management in rehabilitation.," *Archives of physical medicine and rehabilitation*, vol. 67, no. 6, pp. 375–379, 1986.

[370] R. Liu, Y. Wang, G. I. Newman, N. V. Thakor, and S. Ying, "EEG classification with a sequential decision-making method in motor imagery BCI," *International journal of neural systems*, vol. 27, no. 08, p. 1750046, 2017.

[371] P. Brunner, L. Bianchi, C. Guger, F. Cincotti, and G. Schalk, "Current trends in hardware and software for brain–computer interfaces (BCIs)," *Journal of neural engineering*, vol. 8, no. 2, p. 025001, 2011.

[372] M. Z. Baig, N. Aslam, and H. P. Shum, "Filtering techniques for channel selection in motor imagery EEG applications: a survey," *Artificial intelligence review*, vol. 53, no. 2, pp. 1207–1232, 2020.

[373] P. Arpaia, F. Donnarumma, A. Esposito, and M. Parvis, "Channel selection for optimal eeg measurement in motor imagery-based brain-computer interfaces," *International Journal of Neural Systems*, vol. 31, no. 03, p. 2150003, 2021.

[374] Q. Gao, X. Duan, and H. Chen, "Evaluation of effective connectivity of motor areas during motor imagery and execution using conditional granger causality," *Neuroimage*, vol. 54, no. 2, pp. 1280–1288, 2011.

[375] U. Halsband and R. K. Lange, "Motor learning in man: a review of functional and clinical studies," *Journal of Physiology-Paris*, vol. 99, no. 4-6, pp. 414–424, 2006.

[376] O. Bai, D. Huang, D.-Y. Fei, and R. Kunz, "Effect of real-time cortical feedback in motor imagery-based mental practice training," *NeuroRehabilitation*, vol. 34, no. 2, pp. 355–363, 2014.

[377] C. Jeunet, C. Vi, D. Spelmezan, B. N'Kaoua, F. Lotte, and S. Subramanian, "Continuous tactile feedback for motor-imagery based brain-computer interaction in a multitasking context," in *IFIP Conference on Human-Computer Interaction*, pp. 488–505, Springer, 2015.

[378] J. R. Wolpaw, D. J. McFarland, G. W. Neat, and C. A. Forneris, "An eeg-based brain-computer interface for cursor control," *Electroencephalography and clinical neurophysiology*, vol. 78, no. 3, pp. 252–259, 1991.

[379] N. Birbaumer, N. Ghanayim, T. Hinterberger, I. Iversen, B. Kotchoubey, A. Kübler, J. Perelmouter, E. Taub, and H. Flor, "A spelling device for the paralysed," *Nature*, vol. 398, no. 6725, pp. 297–298, 1999.

[380] S. Lee, S. Ruiz, A. Caria, R. Veit, N. Birbaumer, and R. Sitaram, "Detection of cerebral reorganization induced by real-time fMRI feedback training of insula activation: a multivariate investigation," *Neurorehabilitation and neural repair*, vol. 25, no. 3, pp. 259–267, 2011.

[381] E. Abdalsalam, M. Z. Yusoff, A. Malik, N. S. Kamel, and D. Mahmoud, "Modulation of sensorimotor rhythms for brain-computer interface using motor imagery with online feedback," *Signal, Image and Video Processing*, vol. 12, no. 3, pp. 557–564, 2018.

[382] C. Neuper, R. Scherer, S. Wriessnegger, and G. Pfurtscheller, "Motor imagery and action observation: modulation of sensorimotor brain rhythms during mental control of a brain–computer interface," *Clinical neurophysiology*, vol. 120, no. 2, pp. 239–247, 2009.

[383] T. Ono, A. Kimura, and J. Ushiba, "Daily training with realistic visual feedback improves reproducibility of event-related desynchronisation following hand motor imagery," *Clinical Neurophysiology*, vol. 124, no. 9, pp. 1779–1786, 2013.

[384] A. Vourvopoulos, A. Ferreira, and S. B. i Badia, "Neurow: an immersive vr environment for motor-imagery training with the use of brain-computer interfaces and vibrotactile feedback," in *International Conference on Physiological Computing Systems*, vol. 2, pp. 43–53, SCITEPRESS, 2016.

[385] H. Ziadeh, D. Gulyas, L. D. Nielsen, S. Lehmann, T. B. Nielsen, T. K. K. Kjeldsen, B. I. Hougaard, M. Jochumsen, and H. Knoche, ""mine works better": Examining the influence of embodiment in virtual reality on the sense of agency during a binary motor imagery task with a brain-computer interface," *Frontiers in psychology*, vol. 12, 2021.

[386] K. A. McCreadie, D. H. Coyle, and G. Prasad, "Is sensorimotor bci performance influenced differently by mono, stereo, or 3-d auditory feedback?," *IEEE Transactions on Neural Systems and Rehabilitation Engineering*, vol. 22, no. 3, pp. 431–440, 2014.

[387] T. Hinterberger, N. Neumann, M. Pham, A. Kübler, A. Grether, N. Hofmayer, B. Wilhelm, H. Flor, and N. Birbaumer, "A multimodal brain-based feedback and communication system," *Experimental brain research*, vol. 154, no. 4, pp. 521–526, 2004.

[388] T. R. Coles, N. W. John, D. Gould, and D. G. Caldwell, "Integrating haptics with augmented reality in a femoral palpation and needle insertion training simulation," *IEEE transactions on haptics*, vol. 4, no. 3, pp. 199–209, 2011.

[389] S. Jeon and S. Choi, "Haptic augmented reality: Taxonomy and an example of stiffness modulation," *Presence: Teleoperators and Virtual Environments*, vol. 18, no. 5, pp. 387–408, 2009.

[390] R. Leeb, K. Gwak, D.-S. Kim, *et al.*, "Freeing the visual channel by exploiting vibrotactile bci feedback," in *2013 35th Annual International Conference of the IEEE Engineering in Medicine and Biology Society (EMBC)*, pp. 3093–3096, IEEE, 2013.

[391] S. Liburkina, A. Vasilyev, L. Yakovlev, S. Y. Gordleeva, and A. Y. Kaplan, "A motor imagery-based brain–computer interface with vibrotactile stimuli," *Neuroscience and Behavioral Physiology*, vol. 48, no. 9, pp. 1067–1077, 2018.

[392] A. Chatterjee, V. Aggarwal, A. Ramos, S. Acharya, and N. V. Thakor, "A brain-computer interface with vibrotactile biofeedback for haptic information," *Journal of neuroengineering and rehabilitation*, vol. 4, no. 1, pp. 1–12, 2007.

[393] L. Pillette, B. N'kaoua, R. Sabau, B. Glize, and F. Lotte, "Multi-session influence of two modalities of feedback and their order of presentation on

MI-BCI user training," *Multimodal Technologies and Interaction*, vol. 5, no. 3, p. 12, 2021.

[394] M. Fleury, G. Lioi, C. Barillot, and A. Lécuyer, "A survey on the use of haptic feedback for brain-computer interfaces and neurofeedback," 2020.

[395] "Compute and visualize ERDS maps." `https://mne.tools/stable/auto_examples/time_frequency/time_frequency_erds.html`. [Online; accessed 06-May-2022].

[396] N. Bagh and M. R. Reddy, "Hilbert transform-based event-related patterns for motor imagery brain computer interface," *Biomedical Signal Processing and Control*, vol. 62, p. 102020, 2020.

[397] J. Kalcher and G. Pfurtscheller, "Discrimination between phase-locked and non-phase-locked event-related EEG activity," *Electroencephalography and clinical neurophysiology*, vol. 94, no. 5, pp. 381–384, 1995.

[398] S. Lemm, K.-R. Müller, and G. Curio, "A generalized framework for quantifying the dynamics of EEG event-related desynchronization," *PLoS Comput Biol*, vol. 5, no. 8, p. e1000453, 2009.

[399] M. Engin, T. Dalbastı, M. Güldüren, E. Davaslı, and E. Z. Engin, "A prototype portable system for EEG measurements," *Measurement*, vol. 40, no. 9-10, pp. 936–942, 2007.

[400] E. Erkan and I. Kurnaz, "A study on the effect of psychophysiological signal features on classification methods," *Measurement*, vol. 101, pp. 45–52, 2017.

[401] P. Arpaia, F. Donnarumma, A. Esposito, and M. Parvis, "Channel selection for optimal eeg measurement in motor imagery-based brain-computer interfaces.," *International Journal of Neural Systems*, pp. 2150003–2150003, 2020.

[402] C. Zuo, J. Jin, R. Xu, L. Wu, C. Liu, Y. Miao, and X. Wang, "Cluster decomposing and multi-objective optimization based-ensemble learning framework for motor imagery-based brain–computer interfaces," *Journal of Neural Engineering*, vol. 18, no. 2, p. 026018, 2021.

[403] T.-j. Luo, C.-l. Zhou, and F. Chao, "Exploring spatial-frequency-sequential relationships for motor imagery classification with recurrent neural network," *BMC bioinformatics*, vol. 19, no. 1, pp. 1–18, 2018.

[404] D. Thiyam and E. Rajkumar, "Common Spatial Pattern Algorithm Based Signal Processing Techniques for Classification of Motor Imagery Movements: A Mini Review," *IJCTA*, vol. 9, no. 36, pp. 53–65, 2016.

[405] J. Müller-Gerking, G. Pfurtscheller, and H. Flyvbjerg, "Designing optimal spatial filters for single-trial EEG classification in a movement task," *Clinical neurophysiology*, vol. 110, no. 5, pp. 787–798, 1999.

[406] K. K. Ang, Z. Y. Chin, H. Zhang, and C. Guan, "Mutual information-based selection of optimal spatial–temporal patterns for single-trial EEG-based BCIs," *Pattern Recognition*, vol. 45, no. 6, pp. 2137–2144, 2012.

[407] B. B. C. I. (BBCI), "Dataset iiia: 4-class eeg data," 2006.

[408] R. Leeb, C. Brunner, G. Müller-Putz, A. Schlögl, and G. Pfurtscheller, "Bci competition 2008–graz data set b," *Graz University of Technology, Austria*, pp. 1–6, 2008.

[409] A. Schlogl, J. Kronegg, J. Huggins, and S. Mason, "19 evaluation criteria for bci research," *Toward brain-computer interfacing*, 2007.

[410] A. Gramfort, M. Luessi, E. Larson, D. A. Engemann, D. Strohmeier, C. Brodbeck, R. Goj, M. Jas, T. Brooks, L. Parkkonen, and M. S. Hämäläinen, "MEG and EEG data analysis with MNE-Python," *Frontiers in Neuroscience*, vol. 7, no. 267, pp. 1–13, 2013.

[411] B. Graimann, J. E. Huggins, S. P. Levine, and G. Pfurtscheller, "Visualization of significant ERD/ERS patterns in multichannel EEG and ECOG data," *Clinical neurophysiology*, vol. 113, no. 1, pp. 43–47, 2002.

[412] M. Z. Baig, N. Aslam, and H. P. Shum, "Filtering techniques for channel selection in motor imagery EEG applications: a survey," *Artificial intelligence review*, vol. 53, pp. 1207–1232, 2020.

[413] L. Faes, G. Nollo, and A. Porta, "Information-based detection of nonlinear Granger causality in multivariate processes via a nonuniform embedding technique," *Physical Review E*, vol. 83, no. 5, p. 051112, 2011.

[414] D. C. Montgomery and G. C. Runger, *Applied statistics and probability for engineers*. John Wiley & Sons, 2010.

[415] A. Ghaemi, E. Rashedi, A. M. Pourrahimi, M. Kamandar, and F. Rahdari, "Automatic channel selection in EEG signals for classification of left or right hand movement in Brain Computer Interfaces using improved binary gravitation search algorithm," *Biomedical Signal Processing and Control*, vol. 33, pp. 109–118, 2017.

[416] P. K. Parashiva and A. Vinod, "A New Channel Selection Method using Autoencoder for Motor Imagery based Brain Computer Interface," in *2019 IEEE International Conference on Systems, Man and Cybernetics (SMC)*, pp. 3641–3646, IEEE, 2019.

[417] J. Jin, Y. Miao, I. Daly, C. Zuo, D. Hu, and A. Cichocki, "Correlation-based channel selection and regularized feature optimization for MI-based BCI," *Neural Networks*, vol. 118, pp. 262–270, 2019.

[418] H. H. Ehrsson, S. Geyer, and E. Naito, "Imagery of voluntary movement of fingers, toes, and tongue activates corresponding body-part-specific motor representations," *Journal of neurophysiology*, vol. 90, no. 5, pp. 3304–3316, 2003.

[419] A. M. Batula, J. A. Mark, Y. E. Kim, and H. Ayaz, "Comparison of brain activation during motor imagery and motor movement using fNIRS," *Computational intelligence and neuroscience*, vol. 2017, 2017.

[420] T. Sollfrank, A. Ramsay, S. Perdikis, J. Williamson, R. Murray-Smith, R. Leeb, J. Millán, and A. Kübler, "The effect of multimodal and enriched feedback on smr-bci performance," *Clinical Neurophysiology*, vol. 127, no. 1, pp. 490–498, 2016.

[421] M. A. Khan, R. Das, H. K. Iversen, and S. Puthusserypady, "Review on motor imagery based bci systems for upper limb post-stroke neurorehabilitation: From designing to application," *Computers in Biology and Medicine*, p. 103843, 2020.

[422] P. Arpaia, A. Esposito, F. Mancino, N. Moccaldi, and A. Natalizio, "Active and passive brain-computer interfaces integrated with extended reality for applications in health 4.0," in *International Conference on Augmented Reality, Virtual Reality and Computer Graphics*, pp. 392–405, Springer, 2021.

[423] bHaptics Inc., "Tactsuit with forty vibration motors." `https://www.bhaptics.com/tactsuit/tactsuit-x40`.

[424] H. Cho, M. Ahn, S. Ahn, M. Kwon, and S. C. Jun, "Eeg datasets for motor imagery brain–computer interface," *GigaScience*, vol. 6, no. 7, p. gix034, 2017.

Index

Alpha rhythms, 20
Analog Digital Converter, 10, 67, 120
Artifact removal, 11
    Independent Component Analysis, 61, 124
Attention in rehabilitation, 102

Beta rhythms, 20
Brain-Computer Interface, 3
    active, 7, 23, 161, 197
    asynchronous, 7
    dependent, 7
    endogenous, 7
    exogenous, 7
    invasive, 5
    low cost, 54
    non-invasive, 5, 53
    partially invasive, 5
    passive, 7, 23, 101
    portable, 53
    reactive, 7, 23, 51
    synchronous, 7
    wearable, 53, 61

Classification, 13
Classification accuracy, 14, 78, 126
Classification latency, 78
Classifiers
    Artificial Neural Networks, 14, 126
    Decision Trees, 13
    k-Nearest Neighbors, 13, 126
    Linear Discriminant Analysis, 13
    Linear discriminant Analysis, 127
    Naive Bayes, 14
    Naive-Bayesian Parzen-Window, 177
    Random Forest, 14
    Support Vector Machine, 14, 127
Common Spatial Pattern, 125, 174
Computed Tomography, 18
Correlation-based algorithm, 69
Cross Validation, 14

Delta rhythms, 20
Domain adaptation, 125

Electrodes, 9, 29
    active, 9, 57
    dry, 9, 19, 34, 57
    non-polarizable, 32
    passive, 9
    polarizable, 32
    wet, 9, 19, 29, 33
Electroencephalography, 16
Emotion in human-machine interaction, 103
Engagement in learning, 106
Engagement in rehabilitation, 105
Event-Related Desynchronization, 22, 164, 165
Event-Related Synchronization, 22, 164, 165
Evoked Potentials, 21, 51
    Visually-Evoked Potentials, 21
        Code-modulated Visually-Evoked Potentials, 21
        Steady-State

Visually-Evoked Potentials, 7, 21, 51, 52
Transient Visually-Evoked Potentials, 21

Features extraction, 11, 124
Features selection, 12
Functional magnetic resonance imaging, 17
Functional near-infrared spectroscopy, 17

Gamma rhythms, 20

Hyperparameters, 126

Information Transfer Rate, 4, 79
International 10–20 system, 9

Magnetoencephalography, 17
Motor Imagery, 162
Mutual Information-based Best Individual Feature, 176

Neurofeedback, 163

Oversampling procedure, 120, 125
- ADASYN, 126
- BorderlineSMOTE, 126
- KMeansSMOTE, 126
- SVMSMOTE, 126
- Synthetic Minority Over-sampling Technique, 126

Positron Emission Tomography, 18
Power Spectral Density, 68
Principal Component Analysis, 13, 125

Reproducibility, 54, 108, 115, 117, 139

Sensitivity, 134
Sensorimotor rhythms, 162
Signal amplification, 9, 37
Signal filtering, 9, 11, 120, 124
- Butterworth filter, 67
- Filter Bank, 173
- FIR, 67

Single-Photon Emission Computed Tomography, 18
Smart Glasses, 61
Stress in neuromarketing, 104

Theta rhythms, 20
Transfer Component Analysis, 125
Transmission unit
- Bluetooth, 28, 172
- Wireless, 120, 132

9781032200859